ALLEN COUNTY PUBLIC LIBRARY

3 1833 02719 9329

P9-EEC-372

j372.7
Griffiths, Rachel
Books you can count on
Parent/Teacher

BOOKS YOU CAN COUNT ON

Linking Mathematics and Literature

ALLEN COUNTY PUBLIC LIBRARY

FORT WAYNE, INDIANA 46802

You may return this book to any location of
the Allen County Public Library.

DEMCO

BOOKS YOU CAN COUNT ON

Linking Mathematics and Literature

RACHEL GRIFFITHS   •   MARGARET CLYNE

HEINEMANN
PORTSMOUTH, NH

To John, Deryn, Morwenna and Tim — R.G.
To Leigh and my three girls, Eleanor, Margaret and Regena — M.C.

Allen County Public Library
900 Webster Street
PO Box 2270
Fort Wayne, IN 46801-2270

HEINEMANN EDUCATIONAL BOOKS, INC.
361 Hanover Street   Portsmouth, NH 03801
Offices and agents throughout the world.

First published in 1988 by
THOMAS NELSON AUSTRALIA
480 La Trobe Street
Melbourne Victoria 3000

First published in the United States in 1991 by Heinemann.
91 92 93 94 95   10 9 8 7 6 5 4 3 2

© Rachel Griffiths & Margaret Clyne, 1988

All rights reserved. No part of this publication may be reproduced in
any form without prior permission of the publisher.

Library of Congress Cataloging-in-Publication Data
Griffiths, Rachel
    Books you can count on : linking mathematics and literature   /
Rachel Griffiths, Margaret Clyne.
        p.   cm.
    Reprint. Originally published : Melbourne, VIC. : Thomas Nelson
Australia, 1988.
    Includes bibliographical references and index.
    ISBN 0-435-08322-8
    1. Mathematics -- Study and teaching (Elementary) 2. Literature in
mathematics education   I. Clyne, Margaret.   II. Title
QA135.5.G69   1991   91-6662
372.7 -- dc20   CIP

Designed by Karen Koljo
Cover illustration based on work
by Cooinda Primary School children
Typeset by Post Typesetters
Printed in the United States of America

CONTENTS

OVERVIEW OF MATHEMATICAL TOPICS

The chart which follows is a guide to the mathematical topics which can be found in the thirty-three books and seven poems treated in Chapters 4 and 5.

Problem solving is not included as a heading; we believe that a problem solving approach is implicit in all these activities. For the same reason, language as such is not given a heading: talk and discussion are an important part of all the work in this book. We have however included as a heading 'Recording', to enable teachers to see the range of possible representations and recordings which children may use, such as story or report writing, equations, graphs, drawings and models. The main focus for each activity is shown in the chart in bold print to distinguish it from other mathematics which may be included.

OVERVIEW CHART

	Title	Level	Counting computation	Pattern and order classification	Spatial relations	Measurement, time, money	Recording
1	Ten, nine, eight	L	**Backwards 10–1**				**Writing**
2	The very hungry caterpillar	L	**1–15**, estimating			**Days of week**, fractions of page	Concrete and pictorial representations, writing
3	Teddybears go shopping	L	**Counting**		**Mapping shapes**	**Money, mass, volume, time**	**Variety of recordings and reports**
4	When we went to the park	L	**1–55**, large numbers	Sequencing			**Writing**, concrete representations
5	1 hunter	L	**1–55**	Classifying			Concrete representations, writing, drawing
6	A lost button/ Buttons	L	**Counting** 1–9	**Classifying**			**Equations**, oral reports
7	Goldilocks and the three bears	L	**1:1 correspondence** estimating	**Order, sequencing**	Construction	**Size**, cooking	Graphing, drawing
8	A lion in the night	LM			**Directional vocabulary, mapping**	Money	**Map drawing**
9	A bag full of pups	LM	**Subtraction, 1–12**				Writing, variety or recordings, **equations**
10	Changes, changes	LM		**Sequencing**	3D shapes, construction	Sequencing	Construction, **story telling** and writing
11	The doorbell rang	LM	**Division**, 1–100				Writing, variety of recordings
12	Happy birthday Sam	LM				**Size — height, time — age**	Graphing, variety of recordings
13	When the king rides by	LM	**1–55**	Ordering, **classifying**			Writing, drawing, listing
14	Grandma goes shopping	M		**Classifying**			Listing, table
15	The shopping basket	M	**Counting 1–21, subtraction**	**Triangular numbers, sequencing**	**Location**, mapping	Money	**Map drawing**, listing

Key to levels: L = lower primary, 5 to 7 years of age; M = middle primary, about 7 to 9; U = upper primary, 9 to 12.

OVERVIEW CHART

	Title	Level	Counting computation	Pattern and order classification	Spatial relations	Measurement, time, money	Recording
16	The very busy spider	M		**Spatial patterns**	Shape (web)	**Size, time**	Graphing, variety of recordings, construction
17	Phoebe and the hot water bottles	M	**Counting** 1–158, **grouping**	Grouping		Time — years, volume, capacity	**Graphing**, variety of recordings
18	Ten apples up on top	M	**Counting, addition, subtraction**				**Equations**, variety of recordings
19	Alexander, who used to be rich last Sunday	M	Addition and subtraction (money)			**Money**	**Writing**
20	Dad's diet	MU	**Fractions, ratio**			**Mass**	Graphing, variety of recordings
21	The Hilton hen house	MU	**Counting, addition**		**Plans and elevations**	Scale drawing	Drawing, plan drawing
22	A day on The Avenue	MU		Sequencing		Time — day	**Drawing**, writing
23	A pet for Mrs Arbuckle	MU	Addition (money, distance)	**Classification**	**Mapping**	Money, timetables, distance	Mapping, report writing
24	Something absolutely enormous	MU	**Large numbers** (1000)			**Size**, length, area volume	Writing, variety of representations, **creating and solving problems**
25	The twelve days of Christmas	MU	**Counting, addition** 1–78, 1–364	Triangular numbers, classification		Measuring squares, time — calendar	Writing, drawing, variety of representations
26	All in a day	U			Mapping — globe and atlas	**Time — seasons, clock**	Drawing, writing
27	Anno's mysterious multiplying jar	U	**Multiplication, large numbers**, estimating	**Number patterns — factorials**			Writing
28	Round trip	U			**Shape**		**Collage, writing**
29	King Kaid of India	U	**Doubling, large numbers**, estimating	**Doubling**		**Volume, mass, money**	Variety of recordings, report writing

	Title	Level	Counting computation	Pattern and order classification	Spatial relations	Measurement, time, money	Recording
30	*Mr Archimedes' bath*	LMU				**Volume, displacement,** height, estimating	Graphing, report writing, story writing
31	*Who sank the boat?*	LMU		**Order**, sequence		**Capacity, mass, balance**	Drawing, variety of recordings, report writing
32	*Anno's counting book*	LMU	**Counting 1–12, cardinal, ordinal**	Sequencing, classifying	Picture map	**Time**	Drawing
33	*Sizes*	LMU		**Classifying, similarities, differences**		**Size comparison**	Drawing, writing
34	'Hands'/'Measuring'	L				**Length — informal units**	Variety of recordings
35	'Autumn'/'Apples'	LM	**Subtraction 1–20**			**Equations**, variety of recordings
36	'A dreadful thought'/'A spider's bedsocks'	LM	**Multiplication, 1:1 correspondence,** grouping				Drawing, writing
37	'Mice'	LM	Counting	**Ordering information**			Variety of recordings, drawing, **modelling to solve problems**
38	'Band-aids'	LM	**Counting 1–100, addition, subtraction, re-grouping**				Variety of recordings
39	'Age'	MU	**Addition, subtraction, multiplication, division**			Time — age, dates	Writing, **creating equations and problems**
40	'The song of the shapes'	U			**2D and 3D shapes**		**Writing**

1:

WHY MATHEMATICS AND LITERATURE?

Words and numbers are of equal value, for, in the cloak of knowledge, one is warp and the other woof. It is no more important to count the sands than it is to name the stars.

(The phantom tollbooth, p.67)

MATHEMATICS AND UNDERSTANDING THE WORLD

Mathematics and literature... why the juxtaposition? Don't they occupy opposing camps, different and distinct spots on the time-table, have different aims and objectives? Isn't mathematics dry topics such as long multiplication, simultaneous equations and right-angled triangles, whereas literature orders life experiences and human relationships?

This view of mathematics is common among both children and adults; but to regard mathematics as a set of abstract concepts and skills is as narrow as regarding literature as a set of grammatical constructions and vocabulary. Rather, we should look on mathematics as a means of understanding the world around us: of dealing with information about space, shape, time and quantity by grouping, ordering and transforming.

Mathematics serves to help us understand and live in our world. There is indeed a discipline of mathematics, which includes concepts and skills as well as the ability to think in a mathematical way. However, to impose the abstractions of this discipline on children before they have had sufficient concrete experiences, discussion and opportunities to make generalisations not only proves useless but can prevent children's future development of understanding. The skills and concepts which we hope to teach need to be embedded in a context which provides meaning and purpose for the children.

MATHEMATICS IN TRADITION AND CULTURE

Historically, mathematics has always had a place in the culture and tradition of societies. The Greeks explored mathematical ideas such as infinitesimals through paradoxical stories such as 'Achilles and the Tortoise'. If Achilles travels 10 times as fast as the tortoise, and gives the tortoise a head start, will he ever over-take the tortoise? If Achilles gives the tortoise 100 metres' start, then by the time Achilles has made up that 100 metres, the tortoise is 10 metres ahead. When Achilles has travelled the extra 10 metres, the tortoise is still 1 metre ahead.

And so on (Kline: 1964, p.403, Northrop: 1963, p.120).

For the Greeks, geometry described the universe, both physically and metaphysically. The four elements, earth, air, fire and water, were themselves composed of fundamental particles in the shape of four of the regular polyhedra (fire: tetrahedron, air: octahedron, water: icosahedron and earth: cube) whilst the fifth shape, the dodecahedron, God reserved for the shape of the universe itself (Kline: 1964, p.47).

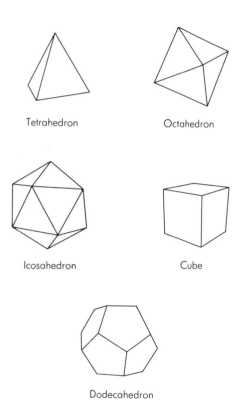

Tetrahedron   Octahedron

Icosahedron   Cube

Dodecahedron

Numerology was an important part of many cultures and was used extensively to interpret and create texts. Composers such as J. S. Bach and Bela Bartok incorporated numerology into their musical scores. The I Ching, the classic Chinese book of wisdom, is based on numbers derived from the throwing of sticks or coins. Certain numbers also seem to have mystical and religious qualities — especially 3, 7 and 12 which appear in the Bible, fairy stories, folk tales and legends. Superstitions about numbers survive to this day: a house rented by one of the authors was numbered 11a to avoid the ill-luck traditionally associated with 13. Then

there are old riddles, number rhymes and chants such as:

> Two legs sat upon three legs,
> With one leg in his lap;
> In comes four legs
> And runs away with one leg;
> Up jumps two legs,
> Catches up three legs,
> Throws it after four legs,
> And makes him bring back one leg.

Consider also the famous Riddle of the Sphinx:
> What is it that goes on four legs in the morning, two legs at noon, and three legs in the evening (Bryant: 1983, p.163)?

This riddle is well known:
> As I was going to St Ives
> I met a man with seven wives.
> Each wife had seven sacks,
> Each sack had seven cats,
> Each cat had seven kits.
> Kits, cats, sacks and wives,
> How many were going to St Ives?

It bears a remarkable resemblance to problem 79 in the Rhind papyrus, written about 1650 BC, which concerns:
> 7 houses
> 49 cats
> 343 mice
> 2401 spelt (grains)
> 16807 hekat (Wells: 1986, p.71)

The mediaeval Arab poet al-Sabhadi, in writing of the invention of chess in India, told the story of the inventor who demanded as his reward as many grains of wheat as it would take to fill the sixty-four squares of the chessboard if the first square held one grain, the second two, the third four, the fourth eight, and so on, doubling the number of grains from one square to the next. The king grants this request, and is then dismayed to find that all the granaries of all the kingdoms on earth cannot supply that much wheat. Activities based on this story can be found on page 48. Al-Sabhadi's version of this story contains a neat twist at the end: since it is impossible for the king to pay his debt, his reckoner advises that the inventor must count out the grains himself (Ifrah, 1987, pp.434–7).

In more recent times, an Oxford mathematician incorporated many mathematical concepts and paradoxes into his books for children, particularly *Alice in Wonderland* and *Through the looking glass*. Lewis Carroll's particular interest was logic and language:

> 'Take some more tea,' the March Hare said to Alice, very earnestly.
>
> 'I've had nothing yet,' Alice replied in an offended tone, 'so I can't take more'.
>
> 'You mean you can't take *less*,' said the Hatter, 'it's very easy to take *more* than nothing'. (Carroll: 1974, p.109)

Mathematics and language are, as Carroll demonstrates in this account, inextricably linked, yet at the same time the dialogue highlights the difference between everyday language and mathematical usage.

MATHEMATICS AND LANGUAGE

Children learn mathematics through using language, therefore opportunities for discussion during all stages of mathematical learning are important. Children develop mathematical concepts through the use of informal language, and move gradually towards a formal terminology and a symbolic method of recording.

Many researchers have documented the importance of young children having experience with informal language and concrete representations before using the more formal language of mathematics (Clements: 1982, Labinowicz: 1985, Irons: 1985, Hughes: 1986). The combination of mathematics and literature, used in conjunction with opportunities for talk and discussion, allows children to grapple with mathematical concepts in a meaningful context.

MATHEMATICS AND CONTEXT

The importance of context in establishing mathematical thinking has been emphasised by researchers such as Margaret Donaldson, James McGarrigle and Martin Hughes in Edinburgh (Donaldson: 1979, Hughes: 1986). By placing tasks in a context which was familiar and which made sense to the children, they found that young children were able to perform at levels which Piaget, using similar tasks but set in unfamiliar contexts, claimed they were unable to achieve. For instance, Piaget set up a model of three mountains and asked children to identify from a set of pictures the viewpoint of a doll at a different position from the child. He found that children up to the age of 8 or 9 were unable to do this successfully. Hughes adapted the task so that the child had to hide a 'boy' doll from two 'policeman' dolls. In this case, children as young as 3½-5 years were able to co-ordinate the points of view of the two policemen and hide the boy correctly. Not only was the situation more familiar to the children (hiding, though not usually from policemen, is a natural activity for most children) but there was also a purpose to the activity (Donaldson: 1979, pp.20-2).

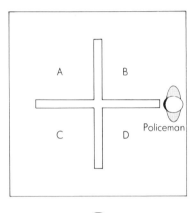

Child

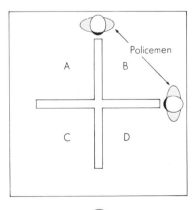

Child

Another well-known Piagetian task tests class inclusion by presenting the child with a collection of, for example, plastic beads, some of which are red, and a smaller number of which are white, and asking the child whether there are more red beads or more plastic beads. Typically, children compare the two subsets (red beads and white beads), rather than comparing the whole set with the subset, and the usual answer from a five-year-old looking at four red beads and two white beads is that there are more red beads than plastic beads.

McGarrigle suspected that it was the unfamiliarity of the situation and of the language rather than the class inclusion itself which was confusing the child. After all, most adults who are unfamiliar with this kind of experiment will need to think twice to be sure that they know what they are being asked. He therefore devised a context which was more meaningful for children; he set up a teddy, a toy chair and a toy table in a line, with four counters ('steps') to form a path from the teddy to the chair and two more counters on to the table. He then asked the question: are there more steps to go to the chair or more steps to go to the table?

He found that this question was answered correctly by the majority of children. (Donaldson: 1979, pp. 45-6) It seems that this very simple story setting provided a context in which the question made sense to the children, and in which they were therefore able to manipulate the concepts.

Literature goes beyond the story setting and provides a context which is interesting and meaningful to children, as well as presenting them with investigations which interest and excite them.

MATHEMATICS AND LITERATURE

Mathematical ideas and concerns are present in literature of all kinds today, and indeed the purposes or functions of mathematics and literature are closer than might at first appear. One function of mathematics is to order the world around us. So does literature. Mathematics is concerned with classification. So is literature. Mathematics is concerned with problem solving. So is literature. Mathematics looks at relationships. So does literature. Mathematics involves patterns. So does literature. And mathematics and literature both have aesthetic appeal. Without taking this analogy too far, we contend that mathematics and literature have strong links, both in content and in structure, and that these links should be explored to make more effective the understanding of both mathematics and literature.

2:

BENEFITS OF USING LITERATURE IN MATHEMATICS

'A light meal it shall be', roared the bug...
When [the king] lifted the covers, shafts of
brilliant-coloured light leaped from the
plates and bounced around the ceiling, the
walls, across the floor, and out the
windows.

(The phantom tollbooth, p.74)

STORY AND MATHEMATICS

Mathematics and literature have strong links and the benefits of using them together are many. Literature has an aesthetic and universal appeal to both adults and children. The magic of telling and reading stories should be offered to all children. Most children beginning school have already had some experience with books and this has assisted in shaping their perceptions of the world. Books extend and develop children's ideas of the world, but at the same time these ideas are bounded by the confines and constraints of the story. Activities suggested here will simultaneously allow for mathematical exploration and extend and develop the ideas in the story. At the same time the limits of the story can help children to focus on the mathematical ideas within the text.

The familiarity of the book or the story gives children a structure within which to explore mathematics. Such a structure provides children with a defined context within which they can manipulate and develop mathematical concepts. Within various texts and illustrations (do not forget the illustrations as they often carry the mathematical ideas) there are opportunities to involve children in problem solving, pattern and order activities, and classification, as well as other mathematical skills. Children are also provided with experiences which demonstrate that mathematics and literature are interrelated and not separate entities.

Indeed, Kieran Egan in *Teaching as story-telling* makes a strong case for the story-form as the basis of teaching in all subjects (Egan: 1989). Tying mathematics to stories humanises the activity and also gives purpose and meaning to mathematics for both teacher and children. Literature then can provide a link between the complexity of the world around us and the highly structured discipline of mathematics.

INTEGRATED STUDIES

Bringing mathematics and literature together also assists the teacher in integrating the curriculum. This approach inevitably includes other curriculum areas such as science (e.g. through *Mr Archimedes' bath*), social education (e.g. through *A day on The Avenue*), and physical education (e.g. through *A lion in the night*). Visual arts and the language arts are also incorporated in all of the work described here.

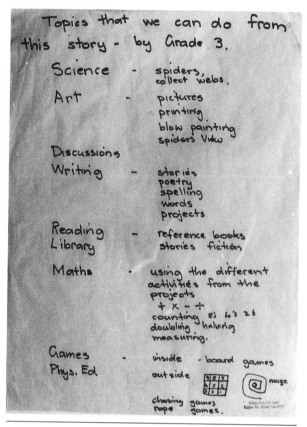

Grade 3 children at Wonga Park did maths projects based on *The very busy spider* and then extended into other areas of the curriculum.

Mathematics should not be imposed upon a work of literature, as this would defeat the purpose of integrating the subject areas: but rather, the mathematics should flow from, and be a natural part of, the book. Using literature in this way needs care, therefore we stress that the activities which may follow a reading or sharing of a book must be a natural development from the text.

SHARED EXPERIENCE

The shared experience which occurs in the initial stages of reading books to children is valuable not only in developing story structure and encouraging various responses to literature but also in developing mathematical thinking. Some of these books are available in large format (big books) which allows a more intimate interaction with the text. Children as a group can share and discuss the mathematical ideas and the language they contain, which can only enrich both aspects of the curriculum.

The sessions we outline later demonstrate how groups of children may share the experience of reading and decision making as a whole class, or as a small group. The class sharing followed by children working and discussing in small groups or pairs continues this social interaction and allows children to build on ideas already suggested. This assists children in clarifying their thinking and in developing concepts and skills. Other activities may have children working in pairs or alone, allowing them time to absorb and think through their task.

LANGUAGE AND MATHEMATICS

The language of mathematics is a stumbling block to many students. Children throughout their mathematical education need to be given time in situations which allow them to use language to describe, explain, report, investigate and question. Practice in using mathematical language assists children in developing and refining their understandings.

Bridging the gap between informal oral language and the formal symbolic code of mathematics is an important but hazardous area. As Lewis Carroll noted, even a simple word such as 'more' is used differently in everyday situations and in a mathematical context.

For children to appreciate the purpose and usefulness of precise mathematical language they need to have a purpose for using that language. We do not offer any panacea for this problem, but we do believe that the context of stories and of problems arising from literature provides a meaning and purpose for children's exploration of mathematical language.

Working and discussing in small groups, reporting to the class and justifying and explaining methods and solutions help children to develop vocabulary and expression. Such experiences demonstrate the need for clarity and comprehensibility in discussion. Recording mathematical work for others to read or interpret similarly sets up a need and purpose for clear and effective communication. By listening to children and talking to them about their work the teacher will gain insight into children's mathematical thinking and be in a better position to assist them.

Reading techniques valued in the language area are supported when teachers use literature to explore mathematics. In many books the children have to extract the mathematics from the text, which gives the teacher an opportunity to gauge their comprehension. Sequencing is another activity which supports children's reading development. Also, many activities demand detailed text reading and repeated readings, which gives reading practice with a purpose. Writing is a natural part of the activities we present. Modelled writing by the teacher and the opportunity for innovative text and report writing all contribute to encouraging clarity of expression and the development of imaginative thinking in mathematics and writing.

Whilst developing the mathematics from *Phoebe and the hot water bottles* children also covered these curriculum areas.

ALLOWING FOR A RANGE OF RESPONSES

In the activities to be outlined children are given opportunities for a range of responses. Problems which allow for a variety of methods of attack permit children to respond at their own level of competence (see Johnson: 1989,

pp. 30–1). This helps children to develop ownership and control of their learning and at the same time involves them in decision making. In some activities children may identify the mathematics and devise their own problems and investigations as well as choosing material, methods and presentation.

By experiencing a variety of methods and sharing these with others, children are helped in shaping and refining their own methods. Such experiences will assist children in choosing more appropriate methods and materials in different situations.

The children in grade 3 at Wonga Park made this list of activities for 'Maths Is'. When new words are discovered they are added to the list. This chart is used by the children when they are working on a Maths Project.

The teacher needs to be aware of the needs and range of abilities within the class when using books this way. The open-ended activities allow children to make their own decisions and follow their own investigations, giving them the opportunity to participate and contribute at their own level. Those activities which are more teacher directed may allow for a specific mathematical focus whilst still permitting children to choose their own materials, methods and ways of presentation.

RANGE OF BOOKS AVAILABLE

During the time we have collected books to use in the classroom we have been amazed at the wide range available which have potential for teaching a multitude of mathematical concepts. For example, Pat Hutchins, Mitsumasa Anno and Pamela Allen are just three author-illustrators who present us with new insights into mathematics. *Anno's mysterious multiplying jar* is a story which investigates multiplication and large numbers. Pat Hutchins in *Clocks and more clocks* looks at the concept of time moving on. And Pamela Allen explores volume and displacement in *Mr Archimedes' bath.*

Certainly not all mathematics can or should be taught through literature. Real life experiences, games, manipulation of concrete materials, and appropriate worksheets or pages from a text book all have their place in mathematics programs. However, we do suggest that using literature to teach mathematics is an additional and very useful strategy to add to teachers' repertoires.

3:

DEVELOPING MATHEMATICAL THINKING THROUGH LITERATURE

'Let's travel by miles', advised the
Humbug, 'it's shorter'.
'Let's travel by half inches', suggested Milo,
'it's quicker'.

(The phantom tollbooth, p.145)

Using books, stories and rhymes to stimulate thinking about mathematics and to develop and re-inforce mathematical concepts enhances children's understanding of mathematics, promotes their enjoyment of the subject and develops their conception of mathematics as an integral part of human knowledge.

Children find the exploration of mathematics linked to a picture book intrinsically motivating; and the number of concepts illustrated in such books affords scope for a wide range of activities and investigation.

ILLUSTRATING CONCEPTS

A good example of a picture book is Jan Pienkowski's *Numbers*, which shows the invariance of number by illustrating on the left-hand page sheep in an array, and on the facing page the sheep scattered over the hill. In contrast to this simple example, Mary Dickinson in *Alex's bed* explores the idea of using space in a creative way. Rather than restricting the placing of furniture to a two-dimensional floor plan, she uses the three-dimensional space in the bedroom.

POSING PROBLEMS

Young children can be asked, for example, to find out how many teddy bears are in Susanna Gretz's book *Teddybears 1-10*. From *The doorbell rang* by Pat Hutchins children can solve the problem of sharing out all the cookies baked by Ma and Grandma.

DIRECTING INVESTIGATIONS

Pamela Allen's *Who sank the boat?* can lead into investigation of floating and sinking, displacement, and order of events. The traditional story *King Kaid of India* can lead to the investigation of large numbers, doubling, mass, volume, and much more.

DEVELOPING RECORDING

In *The doorbell rang* by Pat Hutchins children can solve the problem of sharing out the cookies by choosing their own method of recording, just as younger children can when finding how many pieces of fruit the caterpillar ate through in Eric Carle's *The very hungry caterpillar*.

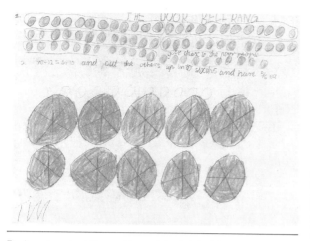

Tim has shared out the cookies and divided the leftovers into sixths.

DEVELOPING MATHEMATICAL LANGUAGE

A lion in the night by Pamela Allen allows children to make their own journey and create their own maps as in the story. This gives children the opportunity to talk and to use the vocabulary already illustrated through the reading and sharing of the text.

STIMULATING INDEPENDENT THINKING

Children can develop their own sequence of events using one set of blocks, in the same way as Pat Hutchins does in *Changes, changes*. *Something absolutely enormous* by Margaret Wild allows children to answer in individual ways the question presented at the end of the book: what will Sally bake, and what problem could it solve?

ENCOURAGING CHOICE AND DECISION MAKING

These activities allow children to choose between a variety of methods. Some books lend themselves to children identifying the problems present in the book and choosing topics and activities. For example, Susanna Gretz's *Teddybears go shopping* includes money, spatial relations, classification and mass. It can be the starting point for an integrated maths unit.

EXTENDING THE PLEASURE OF LITERATURE

Children respond with interest and pleasure to a book they enjoy and the mathematical problems and activities presented or identified may be considered an extension of that pleasure. Children display a high interest in such activities and are strongly motivated to undertake the tasks.

CONCLUSION

We have found that a literature-based approach to mathematics is time-consuming but extremely valuable, both for the involvement of the children and for the work children produce. Teachers have many other pressures on their time, so we recommend that they plan to allocate sufficient time for children to work through a unit. As reading, writing, discussion, art and other curriculum areas are incorporated in these activities, it may not be necessary to run separate reading, writing and mathematics sessions when the class is involved like this.

Experience will assist in choosing appropriate texts for mathematical development as more books are used in this way. Teachers may also discover, as we, and teachers who were involved in the trialling did, that children become keen observers of mathematical ideas and learn to identify areas suitable for mathematical investigation. In the activities which follow we have emphasised those which have a mathematical focus. Teachers should be aware that activities in other curriculum areas can be developed at the same time and that mathematics may be the central focus for a topic (e.g. time in Pat Hutchins' *Clocks and more clocks*) or a starting point for an integrated study (e.g. fairytales — *Goldilocks and the three bears*).

4:

ACTIVITIES BASED

ON INDIVIDUAL

BOOKS

'Why, can you imagine what would
happen if we named all the twos **Henry** or
George or **Robert** or **John** or lots of other
things? You'd have to say **Robert** plus **John**
equals four, and if the four's name were
Albert, things would be hopeless.'

(The phantom tollbooth, p.147)

Children at Cooinda used MAB to make 1000, and created Sally's
absolutely enormous thing.

This chapter demonstrates how you might use
particular books to teach mathematics. However, individual teachers may decide to work
in different ways or to modify our suggestions.
The more you use books in this way, the easier
it becomes for both teacher and children to
think of mathematical problems and activities
based on the books. Some activities will fall
easily into a lesson slot of 45 to 60 minutes,
others will be shorter or longer; some will have
a specific aim or objective, others are more open
to interpretation by teacher and children.

Each plan follows a similar format:

Teacher information

This provides extra information which may be
useful to teachers. It is not included in every
lesson plan.

Synopsis

A brief description of the story, together with
a suggested level, **L** (Lower primary, 5 to 7
years of age), **M** (Middle primary, about 7 to
9) or **U** (Upper primary, 9 to 12 years). Other

books suitable for post-primary use have been marked **PP** in the Bibliography. It is the mathematical activities, not the story, which we are classifying. These levels are suggestions only, but are based on our classroom experiences.

Focus

This is the main mathematical idea, concept or process which the lesson develops, though each lesson will of course include other mathematical ideas and skills. It is likely that some teachers will identify other focuses for some of these books, and will decide to develop those. For further information on the mathematical concept covered, see the Overview Chart on page vii.

Preparation

We assume that paper, pencils, Textas and a variety of concrete materials both structured and unstructured will be readily accessible to the children. Where particular material is needed, we note this. More information on materials is given in Chapter 9.

Activities

We suggest one or more activity for each book. Some activities are written as a sequence, with each change of action signalled by a star; for some books, such as *Something absolutely enormous*, there are a number of alternative activities. In either case, the teacher should interpret the suggestions to suit her or his own group of children.

Before starting any activity, it is an advantage if the children are familiar with the story. Therefore, the teacher may read the book one day and leave it out for children to read and browse through before starting the suggested activities a few days later.

Just as familiarity with the story is important, so is familiarity with the activity. We have found that children frequently want to repeat an activity, perhaps using different materials, and we believe that this contributes greatly to their learning.

An important part of every activity is the discussion of ideas and the sharing of results, whether or not this is explicitly stated. Some

further thoughts on classroom organisation are given in Chapter 9.

Extension

These are additional activities which teachers and children may like to explore. Some follow on from the main activities; others stand on their own.

Further reading

These are books or stories which have a similar focus to the book discussed, or which extend the ideas in the book.

Indicators

The indicators are intended to help teachers to monitor children's mathematical development and to assess their progress. We have focused on what we think are the most important features of the activities. For further information on assessment see Chapter 10.

Length of session

We must emphasise that this is very approximate, and refers only to the main activities for each book. The length of time taken will depend on the responses of the children and the teacher to the book and the mathematics in it. In some cases a break (lunch-time or overnight) may be beneficial to allow children time to reflect on the activity.

TEN, NINE, EIGHT,
Molly Bang

Greenwillow, 1983. **(L)**

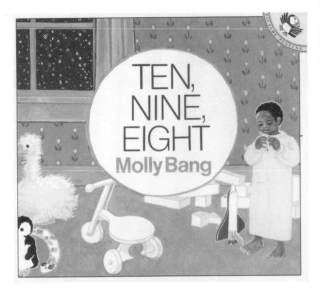

Synopsis

A countdown to bed-time starting with '10 small toes all washed and warm' and ending with '1 big girl all ready for bed'.

Focus

Counting backwards.

Activities

★ Read and discuss the story. Ask the children to predict the number on the following page to help establish the counting pattern.

★ Write a class story using the number pattern.

★ Discuss places around the school with which the children are familiar — the playground, the assembly hall, the classroom. Vote on which place they prefer to write about.

★ Brainstorm and list items that might be found in the place the children have selected. Then go there. Ask the children to identify additional items and to count them in preparation for writing their story. Encourage them to look for the less obvious, such as 'four corners on the jumping mat' or 'six bars on the climbing frame'.

> Ten white mats for rocking and rolling
> Nine bouncing balls spinning on nine fingers
> Eight round hoops for twirling around your waist
> Seven children running fast!
> Six silver stumps make two wickets for playing cricket
> Five red markers mark the boundaries we have to play in
> Four scratch 'n' smell stickers for people who do the right thing—
> chocolate, strawberry, banana and mint
> Three white rings in the target we throw balls at
> Two white basket ball rings with red white and blue zig-zag nets of string
> One big blue fat crash mat to jump and roll on.
>
> Grade One children wrote this after a visit to the gymnasium.

★ After returning to the classroom, write the text as a group. Include appropriate descriptions of each item as in *Ten nine eight*.

★ The text can be illustrated by the children.

Extension

★ The children can write individual books, using the counting back pattern, based on their own bedrooms, houses or somewhere special to them.

★ Write a class book of a countdown to a special event such as Christmas, an excursion or a class party.

> Twelve days of Christmas, my dad sings that to me,
> Eleven days to decorate a sparkling Christmas tree,
> Ten glass balls all glittering and round,
> Nine coloured streamers tumbling to the ground,
> Eight white snowflakes shimmering like ice,
> Seven little ginger cakes, brown, and full of spice,
> Six red Santas, each one fat and jolly,
> Five prickly sprigs, of green and red-bright holly,
> Four golden angels singing loud and clear,
> Three ringing bells, bringing in New Year,
> Two white candles send a flickering light,
> One star near Heaven shining through the night.
> ### MERRY CHRISTMAS
>
> Using counting backwards to count down to an event.

Further reading

See the section on **Counting books**, Chapter 7
Waiting for Sunday Carol Blackburn

Indicators

- Are the children able to recognise the counting pattern in the book?
- Before moving there can the children identify appropriate items in the place they have selected?
- Can they identify appropriate items when at the place?

Length of session

45–60 minutes

THE VERY HUNGRY CATERPILLAR, Eric Carle

Putnam, 1989.   **(L)**

Synopsis

A long-time favourite in which a caterpillar eats through mountains of food, which results in a tummy-ache. The caterpillar finally becomes a butterfly.

Focus

Days of the week, recording, counting, estimating, fractions.

Preparation

- A variety of concrete materials both structured and unstructured
- Plasticine or Playdoh
- A variety of papers (plain, coloured, grid etc.)

Activities

★ Read the book to the children.

★ From the picture, estimate the number of items eaten on Saturday. Count them.

★ Estimate the number of eggs shown on the end papers. Count them.

★ Ask the children to represent/record the number of pieces of fruit the caterpillars ate (remember the watermelon on Saturday) in a variety of ways.

★ Pose the problem: how could we make a book with this pattern of cut pages? Discuss the number of parts, measuring etc. The children could experiment with paper-folding to explore the concept.

★ The children can write a new text as a class, group or individual activity. As a class activity, children may offer suggestions of food or animal. The teacher may need to model the story structure, for example, 'On Monday _____ ate one _____'. Allow the children to work in small groups to make a book.

A very hungry grandpa

1 hungry grandpa went out to afternoon tea and he ate one piece of cake.

Grandpa ate two biscuits.

Grandpa ate three hot cross buns.

Grandpa ate four chocolates and Grandpa ate four apples.

Grandpa ate five cupcakes.

Grandpa ate six teddy bear biscuits.

Grandpa ate seven cakes and was full. He had a nap in the shelter shed.

Preps at Wonga Park wrote this story.

Extension

★ Use the *Butterfly* frieze from the *Informazing* series (Methuen, 1987) to examine the time-line and what happens at each stage.

★ The *Tadpole* frieze from *Informazing* could be a model for the children to use in writing a class book.

Indicators

● Were the children's recordings creative, attractive and clear?
● Could other children interpret them?
● Did the children use a number pattern in their writing?
● When the children wrote their new text, did they use the days in the correct sequence? (It doesn't matter where they started.)

Length of session

At least 60–90 minutes.

Further reading

See sections on **Counting books**, Chapter 7, and on **Time**, Chapter 8.

TEDDYBEARS GO SHOPPING,
Susanna Gretz
Macmillan, 1982.　**(L)**

Teacher information

This activity allows children free choice of topic and material. The teacher should read the book and make equipment accessible to the children.

Synopsis

Four teddybears make a list to go shopping ready for Andrew's return home. They mislay the list and each other. On returning home they find they have plenty of ice-cream but other items are missing. Can you see the shopping list?

Focus

Shopping, identifying mathematics topics, setting and carrying out investigations, reporting.

Preparation

● A wide variety of measuring equipment: e.g. rulers, balances, measuring tapes, grids, counters, MAB, coins.
● A variety of unstructured materials.

Activities

★ Read the book to the children.

★ Ask the children to brainstorm, as a class, all the mathematics they can identify in the book.

In trialling year 1 children identified counting, shapes, weighing, mapping, mazes and money, worked in small groups and developed investigations.

★ The children can then choose an area listed and work in pairs or small groups to develop an activity. Each group is responsible for setting up an investigation for their chosen area, and for reporting back to the class. Each group could be asked to make a display or a written or oral report of their investigation.

Extension

★ Ask the class how much the bears' shopping would cost.

Further reading

Teddybears' moving day, Susanna Gretz (mapping)
Teddybears 1–10, Susanna Gretz (counting)
See the section on **Money**, Chapter 8.

Indicators

● What kind of mathematics did the children identify?
● How successfully did the children work in groups and report to the class?
● Did the children reach the goal they had set themselves?

Length of session

At least 60 minutes, depending on the group of children and the activities chosen.

WHEN WE WENT TO THE PARK, *Shirley Hughes*

Lothrop, 1985.   **(L)**

Synopsis

A counting book with a warm story about going to the park with Grandpa. Each number deals with the observations made in the park and concludes with going home for tea.

Focus

Counting, sequencing.

Activities

★ After reading the book to the children show them the end pages. Discuss the pattern. Discuss the different activities the girl and Grandpa observed. Invite the children to talk about what they do when they go to the park.

★ Ask the children to suggest other places and events they go to, for example: a circus, shopping, the zoo, Moomba, parties, and excursions.

When we went to the circus
we saw one ringmaster
two horses galloping at the circus
three lions going through a hoop
four clowns squirting water
five elephants standing on a stool
six seals balancing balls
seven tigers running
eight acrobats swinging to one another
nine monkeys climbing up a tree
ten ponies walking
and so many people we couldn't believe it!

After writing this story preps at Auburn Primary School investigated the number of animals and the number of people in their text.

★ A story based on these suggestions can then be written using *When we went to the park* as a model. This will probably be most successful as a whole class activity, before inviting the children to work in small groups on their chosen topic. Before writing the whole class story allow the children to vote on the subject matter. During the process of writing give them opportunities to discuss and justify agreed selections.

Extension

★ In the chosen innovative text use size comparison. For example, if the zoo is chosen start with small or large animals and continue to keep the chosen pattern.

---

1 black cat purring because the beard is warm
2 brown puppies playing
3 green frogs hopping along
4 yellow chicks squeaking for food
5 blue lizards are hungry
6 grey snails crawling slowly
7 goldfish swimming in the beard
8 orange caterpillars crawling around
9 grasshoppers jumping for joy in their new home
10 red beetles looking for a home.

Prep. children at Belgrave South read *Creatures in the beard* and combined the ideas from both stories.

★ What do we see so many of at Moomba or when we are shopping that 'we can't count them all'? Make a list with the children:
● stars
● grains of sugar
● hundreds and thousands
● ants
● raindrops

★ Make a class book of *Things We Can't Count Easily*.

Further reading

See the section on **Counting books**, Chapter 7

Indicators

● When writing in groups or in pairs, how did the children interact? Observe how decisions were made.

● Did the children successfully identify suitable items to fit the chosen theme?
● Could they identify items of which there was likely to be only one, or so many you couldn't count them all?

Length of session

60–90 minutes

1 HUNTER,
Pat Hutchins

Greenwillow, 1982.   **(L)**

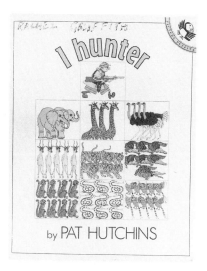

Synopsis

This is a counting book in which the story is carried through the illustrations. The hunter walks through the landscape oblivious of the wildlife, until he turns around to be confronted by all the animals.

Focus

Counting.

Activities

★ After reading and discussing the story, pose the problem: how many animals are there in the story? (The hunter can be included or not as children decide.)

★ Suggest that the children choose more than one material or method to represent the animals.

★ Discuss the results and chosen methods. Which materials or methods do the children prefer? Why?

Extension

★ Children can write their own counting book based on *1 hunter*. This could be done in small groups, or the class could make a big book. The children could then draw, cut out and paste in the appropriate number of each animal. This will probably involve much mathematical thinking and language, as the children decide how many more or fewer of a particular animal are needed.

> We've got too many cats. We need to give three away.

> We need one more horse to make seven

★ The children may classify the animals in the original or in their new text. They should discuss and justify their groupings and their chosen criteria.

Further reading

See the section on **Counting books**, Chapter 7.

Indicators

● What material or method did the children use?
● Were the children systematic in their counting?
● How did they keep track of their progress?

Length of session

45–60 minutes

A LOST BUTTON,
Arnold Lobel

In Frog and Toad are friends, Harper Junior Books, 1983.

BUTTONS
Avelyn Davidson

In Soldiers, Shortland, (Understanding Mathematics), 1984.   **(L)**

Synopsis

In Arnold Lobel's story, Toad and Frog set out on a walk which is spoiled by the loss of Toad's button. The button is found after a search in which all the animals help.

The poem 'Buttons' follows:

Buttons

Buttons in the button box on a rainy day,
Buttons that are fun to sort when it's too wet to play.
Buttons, buttons, buttons,
Yellow, green and blue,
Red buttons, purple buttons,
Come and sort them too.
Big buttons, brass buttons, shiny, round and small,
I would find it difficult to try and count them all.
Buttons, buttons, buttons,
With four holes and with two.
I will sort the buttons and
Make some sets for you.

Avelyn Davidson

Focus

Classifying, counting, equations.

Preparation

● A large quantity of buttons and other materials suitable for sorting and classifying.

Activities

★ Read the rhyme 'Buttons', and leave it on display so that the children can refer to the text or illustration (a big book version is available) for ideas when sorting buttons at this stage or later.

★ Read the story 'A lost button' to the children and have them act it out, paying attention to the language they use when they describe the buttons.

★ Invite the children to choose buttons and describe them in their own way as the story progresses. They may also like to find buttons which match the ones in the story.

★ Ask the children to work in pairs with a small pile of buttons, and make up statements about the different buttons. The teacher could record some of the statements from the groups when they report for discussion.

Grade 2 children at Wonga Park placed the buttons on Toad's jacket and wrote an equation to match the story.

★ The children can count the number of buttons Toad found, draw his jacket and stick on the same buttons as those in the text. They can then create equations about the number of buttons in the story.

Extension

★ Have the children pick up a handful of buttons and estimate the number. They can then count their buttons and record both estimate and count. Doing this a number of times will help the children to develop estimation and counting skills.

★ By comparing their handfuls with other children's, and handfuls of buttons with handfuls of other materials, the children will further develop these skills. Can they work out why their counts may be different? How many buttons can the teacher pick up? Can they find the biggest hand in the school?

★ Introduce other materials for classification.

Further reading
See the section on **Classification**, Chapter 6.

Indicators
- Can the children describe the buttons in the text and match them from the button collection?
- Can the children make equations, and explain their equations, for the number of buttons on the jacket?
- What formal and informal mathematical language do the children use when sorting and describing the buttons?

Length of session
45–60 minutes

GOLDILOCKS AND THE THREE BEARS   (L)

Teacher information
There are many versions of this story, and the children can compare texts and illustrations. The Paul Galdone (World's Work, 1983) illustrations are delightful. There is a big-book version of the song 'When Goldilocks went to the house of the bears' (Ashtons, *Bookshelf* Stage 1). The song has 'big' 'small' and 'tiny' in place of the traditional 'big' 'middle-sized' and 'little'.

Synopsis
Goldilocks visits the house of the three bears; after sampling their porridge, chairs and beds she falls asleep in the little bear's bed.

Focus
Size, order, 1:1 correspondence, sequencing.

Activities
★ Read (or better, tell) the story. In discussion focus on the number and size of bears, bowls, chairs and beds.

★ Ask the children to make groups of three. Does each group have a big, middle-sized and small child? If not, what do they have?

★ Ask the whole class to put themselves in order of size. They could make a graph of the result.

★ Working in co-operative groups of three or four, the children can make their own 'bear family'. As a group, they decide how many bears they want in their family, then they draw their bear family complete with bowls, chairs and beds. Have the children cut out and order their drawings and paste them on paper or card to make a display. They must decide as a group who does what, and end with a single display from the group.

★ In a small group, ask the children to draw pictures of events from the story and put them in order. The children must be prepared to justify their choice of event and their order. They may also want to discuss time.

★ The children can bring their own bears from home to classify, order and measure them. Ask each child to find a bigger bear and a smaller bear than their own. Put these in order and draw them.

★ Finish the unit of work with a display to which other classes, the principal and parents are invited. Use the display of children's work for activities such as estimating the total number of bears in the 'bear families'. Check this. Do the same for the bowls, chairs and beds.

★ Have a 'Teddy Bear Biscuit Feast'. Share out one or more packets of Teddy Bear biscuits. Estimate the number in the packet. Will the second packet have the same number? Check this. Estimate the number each child will receive. Check this by sharing.

Extension

★ A small group of children can construct the bears' house. Make a variety of materials available. The children could also make other items from the story, and some may like to draw a plan of the house.

★ Cook porridge. Measure or weigh the ingredients, time the cooking and note the temperature of the porridge (informally). Is it too hot, too cold or 'just right'?

Further reading

The three billy goats Gruff
A spider's bedsocks, Phil Mena, page 58.
Many books about bears, including the characters Paddington (Michael Bond) and Winnie the Pooh (A. A. Milne).

Indicators

● Are the children able to match bears, bowls, chairs and beds?
● Can the children order their bear family and its belongings correctly?
● Can the children sequence their drawings of the story, and justify their decisions?
● In co-operative work, do all the children participate?
● Do any groups have difficulty reaching agreement?

Length of session

This activity is probably best taken over a number of sessions.

A LION IN THE NIGHT,
Pamela Allen

Putnam, 1986.   **(L, M)**

Synopsis

A lion 'steals' the baby. All set off in hot pursuit, until the lion growls. The baby and the lion occupy the castle and win the game. And what better way to end such an adventure than by having a magnificent breakfast feast? Then the lion, because it is morning, goes home. Look closely at the first and final illustrations.

Focus

Spatial concepts: directional vocabulary, mapping.

Activities

★ Together with the children, list the vocabulary for spatial concepts (e.g. across, over, into) from the book. Invite the children to add additional words (e.g. under, between, around).

★ Look at the map in the book. Discuss the circular arrangement.

★ Then go outside — an area with play equipment works well for this activity. Children travel on a circular route, using the concepts previously listed, and others as they come to mind. They may take paper and pencil to record their route as they move around. They may have different starting points.

★ Back inside, have the children draw their own circular maps of their travels.

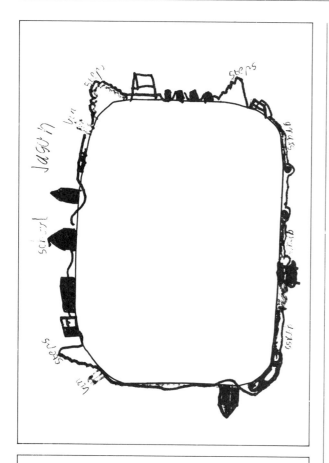

Jason and Aaron

Out the door
around the coana
through the mud
to the bin
up the steps

down the steps
to the Aret room
up the steps
(between the bush its
Jump
over the teabell

over the grass
through the ramback
over the grass

down the grade 4 56
loop - through
the grass

& past The netball
cout
around the coande
up the path to the
bin   up the stbers

through the
door   up the cordoor
in the room

Jason and Aaron (grade 2) recorded their route around Cooinda
Primary School, and Jason drew and labelled the map.

Extension

★ Progress to drawing imaginary maps, or maps of a well-known area (e.g. the school ground or the way to school). Invite the children to invent symbols (e.g. tree, lake). Use other books such as *A pet for Mrs Arbuckle* by Gwenda Smyth, or *Possum Magic* by Mem Fox, to develop more formal mapping skills.

★ How much did the lion's breakfast cost? Use catalogues or research at home to find out.

Further reading

See the section on **Location**, Chapter 6

Indicators

● How accurately are the children using the vocabulary as they move around outside?
● Can the children transfer their outdoor activity onto a map?
● How much detail is evident in their work?
● Can other children follow the map?

Length of session

60–90 minutes or two shorter sessions.

A BAG FULL OF PUPS, D. Gackenbach

Ticknor and Fields, 1983.   **(L, M)**

Synopsis

Mr Mullins has to find homes for twelve pups. He gives them away to people who need them. The last one goes to a small boy who will play with and love the little pup.

Focus

Counting, subtraction, equations.

Preparation

● Write the text onto cards (one for each page) with space for children to attach their representations.
● Display these in order, so the class can see and read the whole story.

Activities

★ Hold a class discussion on the numbers of pups on each page of the story.

★ Ask the children to suggest ways of recording how the number has changed when the first pup is given away. They may suggest drawing 12 pups and crossing one out, tallying, drawing a number line, writing the equation $12 - 1 = 11$, or using concrete material.

★ Repeat this for another page.

★ Then the children can choose a page of text to represent in their own way, and attach their representation to the appropriate page. They may do more than one page, or do more than one representation of the same page.

★ Ask the children to explain and justify their representations. Their equations or diagrams should match the story.

★ Invite the children to write their own stories involving getting rid of or giving away something of their own choice (pets, toys, etc.). Ask them to show what is happening to the numbers at each stage.

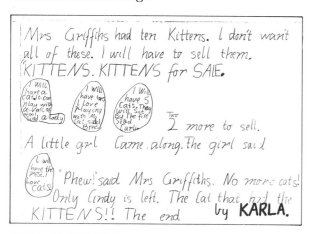

Karla (grade 3 at Mooroolbark) wrote about giving away ten kittens.

Further reading
Five little kookaburras, Brenda Parkes
Five little monkeys (*Bookshelf*, 1986 and Southern Cross, 1987)
Ten loopy caterpillars, Joy Cowley

Indicators
● In writing their new text, did the children use the same or larger numbers?
● What informal and formal mathematical language do the children use?
● What methods of recording are the children comfortable with?

Length of session
If the children write their own stories as well as making representations of this story, 60 minutes or more.

CHANGES, CHANGES,
Pat Hutchins

Macmillan, 1987.   **(L, M)**

Teacher information
If there are insufficient blocks available for the whole class to be involved in the main activity at the same time, some groups may use different construction material or be involved in other activities such as retelling the original story. There is a film version of this book.

Synopsis
This is a no-text book in which a wooden man and woman adapt a set of blocks to suit their needs through a series of changes from a house to a fire engine, to a boat, to a truck, to a train and finally back to a house again.

Focus
Spatial relations (construction), sequencing, story telling.

Preparation
● A set of blocks including a variety of shapes.
● A camera and tape recorder, or video camera, may be useful for recording the children's work.

Activities
★ After showing the book to the children ask them to tell the story. If the children do not

mention it, draw their attention to the fact that the same blocks are used at each stage of the story.

> *The arches are good for doors and windows. If you put them upsidedown they look like wheels.*

> *The flat faces are good for building on top of. The ones with curved edges will roll.*

★ Discuss the shapes of the blocks and their attributes. Ask which blocks are useful for which purposes in building. Have a set of blocks available for children to handle and experiment with.

★ The children may work in pairs to create their own story of 'changes'. They should choose a set of blocks (up to, say, 30) and use the same set to construct a sequence of scenes with a story. The teacher may like to photograph children's constructions and record their stories. Older children may draw and write their stories.

★ The children can read and tell their stories to each other or to the whole class. If photographs have been taken, the children can write down their stories at this stage.

Extension
★ Use different construction material (Lego, Meccano, Stickle bricks, Struts etc.) to make another story of 'Changes'.

Further reading
Danny's dilemma, John Tarlton
See the section on **Shape**, Chapter 6

Indicators
● Are the children competent in their use of blocks for building?
● Do they use appropriate blocks for particular purposes?
● Can the children sequence their story logically?
● Does their story relate to their construction?

Length of session
At least 60 minutes for the construction. A further session will be needed for writing and telling the stories.

THE DOORBELL RANG,
Pat Hutchins

Morrow, 1989.   **(L, M)**

Synopsis

Ma bakes some cookies for Sam and Victoria — six each. But each time they are about to start eating, the doorbell rings, and friends arrive. When there is just one cookie each, the doorbell rings again. It is Grandma, with a huge tray of cookies! Then the doorbell rings...

Focus

Division (sharing).

Activities

★ Pose the problem: how many cookies does each child get when Grandma arrives?

> Every child in the street had some cookies except Mum and grandma because it would ruin their diets. Everyone enjoyed their cookies. I think they wanted some more.
>
> A comment from Camberwell Primary School.

★ With younger children, you may like at first to specify the number of cookies Grandma brings (e.g. 24). By doing this, children with less experience or confidence will be more likely to experience success. This could form the basis for class or group discussion.

★ Children can move on by specifying the number of cookies themselves. Some children will be confident enough to start this way. Older children could count the cookies on the final page, which shows Grandma with her tray of cookies.

★ Ask the children to use two different methods to solve the problem (e.g. concrete material of different kinds, drawing, tallying, role play). The children should explain and justify their solutions. If there are cookies left over (remainders), how do they deal with these?

Extension

★ Have the children continue the story: who arrives when the doorbell rings at the end, and what happens to the cookies?

A problem created by grade 3/4 children at Waverley Park: There are 14 people. Each person got 7½ cookies. Grandma baked 65. How many did the stranger bring?

★ Invite the children to write their own stories involving sharing, and exchange them with other children or groups to solve new problems.

Indicators

● What material or method did the children use?

● Did the children use formal or informal

recordings and language in sharing the cookies?

- How did the children deal with remainders?

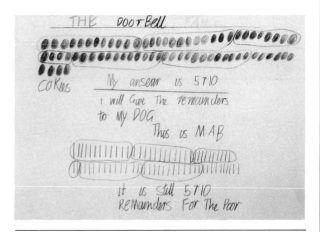

At Camberwell, Aidan used two methods to share out the cookies, and then decided what to do with the leftovers.

Length of session
45–60 minutes to complete the main activity. A second session will probably be needed for the extension.

HAPPY BIRTHDAY SAM, Pat Hutchins

Penguin, 1985.   **(L, M)**

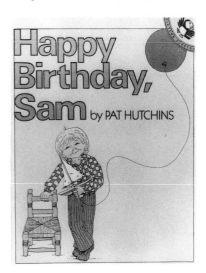

Synopsis
On his birthday Sam is still not tall enough to reach the door handle, light switch or taps, but Grandpa's present solves the problem.

Focus
Size: height, time: age

Activities
★ Read the book to the children. Look back at the first page. Ask whether you really become a whole year older on your birthday.

★ Discuss ages, birthdays and months of birthdays of the children. A class graph can be made of birthdays. In small groups, the children can make statements using information from the graph.

★ Children may like to recall the things they remember they couldn't reach, and how old they were. A class book could be made of *Then and Now*!

★ List things that the children cannot reach now. They can estimate how far up the wall they think they can reach and then check their estimates. How far can the children reach if they stand on a chair? Estimate and find out. Discuss differences in children's reach. Who is taller? Who stood on flat feet and who on tiptoe? Use string, streamers or formal units as appropriate for their stage of development.

Extension

★ Have the children measure the heights and reach of family members and then bring them to school for comparison.

Further reading

Titch, Pat Hutchins

You'll soon grow into them Titch, Pat Hutchins

V.I.P., very important plant, Ted Greenwood

See the section on **Size**, Chapter 6.

Indicators

● Could the children estimate their reach realistically?

● How did the children choose to record their findings?

● Were the children able to extract information from the graph?

Length of session

60–90 minutes.

WHEN THE KING RIDES BY,
Margaret Mahy

Ashton Scholastic (Bookshelf stage 2),
1986   **(L, M)**

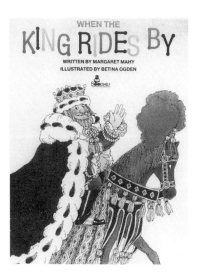

Teacher information

This story is available in big book format.

Synopsis

When the king rides by is a cumulative narrative of a procession headed by the king. It is a counting book in which the increasing numbers are shown in the illustrations, but do not appear in the text.

Focus

Counting, classification.

Activities

★ After reading the book ask the children to look closely at the illustrations. They may take some time to observe how the pages 'fill up'.

★ At this point, re-reading the text and inviting the children to examine each page would be appropriate. Ask the children to relate the text to the illustrations. They will probably join in the counting as you read the text.

★ Invite the children to investigate how many items/people are mentioned in the text. (The class may like to vote to decide whether the king and the soldier with the drum are included.)

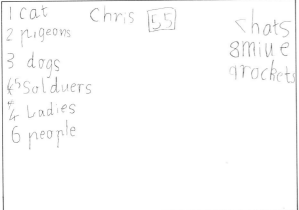

Two different responses from grade 1 children at Glendal to the investigation of the number of items in the text.

★ Ask the children to work in pairs or small groups to classify the items/people which appear in the book. Invite some children to explain and justify their decisions.

Extension

★ A small group may like to present an ordered view of the book — i.e., with the items in an ordered way. Compare Bettina Ogden's illustrations with those in other counting books, for instance Pat Hutchins' *1 hunter*, in which the items are lined up and distinct.

★ The children may work in small groups or as a class to rewrite the text — for example, 'Oh what a fuss when the principal marches by...' (It doesn't have to rhyme.) Moomba could be a good time for this.

Further reading

See the section on **Counting books**, Chapter 7.

Indicators

● How did the children attempt to record and add up the items in the story?
● Were they systematic in their approach?
● Were the children able to justify clearly their decisions on classification?
● Did they compare their decisions with those of other groups?

Length of session

60–90 minutes

GRANDMA GOES SHOPPING,
Ronda and David Armitage

Puffin, 1985.   (M)

Synopsis

This is a book in the traditional style of the 'shopping' game with a rich vocabulary and narrative. Look out for the hidden nursery rhymes and watch Grandma's shopping bag.

Focus

Classification.

Preparation

● A list of the items which Grandma bought for each group.
● Scissors, paste and paper.

Activities

★ Read the book to the class or to a small group. Multiple copies would be an advantage, in order for children to explore the book in groups. Things to look for are: nursery rhymes, Grandma's shopping bag, details in the illustrations, change of style in the artwork and the antics of the alligator and the mouse.

★ After reading the story do oral classification activities with the children, for example: 'All children with runners on go near the door, all children with blue eyes sit on the floor'. Discuss how the classification could be recorded, particularly if some children were in more than one group.

★ Give each group a copy of Grandma's shopping list and ask them to group, classify or sort the items out. Rotate among groups listening to and noting the children's use of language and reasons for categories. The children may need dictionaries to decide groupings if some meanings are unknown.

> A fish for frying can't be in animals because it's dead. It has to be in food.

> A tootling flute and a flugal horn belong with the bicycle built for two because they're both metal.

Grade 3 children at Glendal justify their classifications to the class.

★ Ask each group to report to the class on their categories, and to justify their selection.

Extension

★ The children can write their own shopping lists, and classify their own or other groups' lists.

Further reading

See the section on **Classification**, Chapter 6.

Indicators

● Could the children justify their decisions for categories?
● Did all the children participate in group discussion?
● Did the children know how to make categories and list items?

Length of session

60 minutes.

THE SHOPPING BASKET,
John Burningham

Harper Junior Books, 1980.   **(M)**

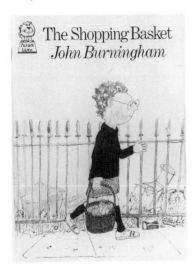

Synopsis

Stephen is sent to do some shopping. He purchases the items correctly, but after some fantastic adventures on the way home the same number are not in the basket. And why did he take so long?

Focus

Counting, subtraction, pattern: triangular numbers, sequencing, spatial relations: location, mapping.

Activities

★ Read the book to the children. Discuss with the children what Stephen is asked to buy.

These children from Camberwell give individual representations of the same neighbourhood.

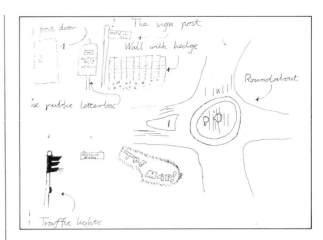

★ Ask the children where their nearest shop is. Do they go there for errands? What do they pass on the way to their local shop? Ask the children to describe and illustrate or map their route to their local shop, as in the second and third pages of the book. Some children may live close to each other and use the same shop. This is a good opportunity for an individual activity followed by comparison and discussion, because the children may be taking similar routes, but making totally different representations of that route.

★ Ask the class: how many things did Stephen have to buy? How many did he bring home? How many did he lose? (They may decide to write equations about the pictures.)

★ Why has he taken so long? Ask what distractions happen when children go to the local shop. (The children may decide that he has broken the egg and eaten the food.)

★ The children can make up their own shopping list using the same (or another) number pattern. They can draw or otherwise represent their shopping.

Extension

★ Discuss what Stephen's shopping would cost. What would their own shopping cost?

Further reading

See the section on **Money**, Chapter 8, and **Location**, Chapter 6.

Indicators

- Did the children draw maps of their route to the shop, or did they just show a sequence of locations?
- Could the children follow each other's 'maps'?
- Did the children produce a variety of number patterns?

Length of session

45–60 minutes.

THE VERY BUSY SPIDER,
Eric Carle

Putnam, 1989.   **(M)**

Teacher information

This book can be used successfully as a stimulus for an integrated curriculum unit on spiders.

Synopsis

The very busy spider completes her web despite repeated interruptions and distractions. The spider, her web and the fly are printed in relief, so this is a book you can feel as well as read.

Focus

Size, time, spatial patterns.

Activities

★ Read the book to the children, and allow them time to read or look at it individually and in small groups. They will enjoy following the progress of the web with their fingers.

★ Invite the children to identify the mathematics in the book. They can then work in groups on an investigation, and present a report and display to the class.

Extension

★ The children can make spiders' webs using string, glue and card.

★ Use the book as a model for writing. Change the setting to the zoo, the desert or the forest.

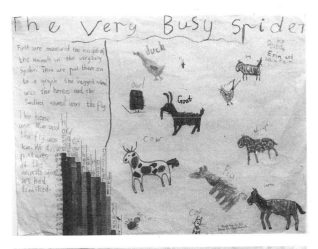

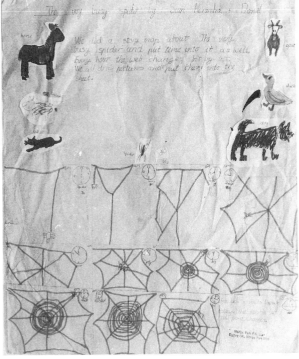

In trialling, grade 3 children at Wonga Park measured the sizes of the animals and graphed them, decided the time at each stage of the story, and investigated the patterns in a spider's web.

Indicators

- Can the children identify the mathematics in the book? In their work, note which skills and concepts they are using.
- Were the children able to work as a group to achieve their goal?

Length of session

Depending on the kinds of investigations undertaken, this could run over several sessions.

PHOEBE AND THE HOT WATER BOTTLES,
Terry Furchgott and Linda Dawson
Picture Lions, 1977.   **(M)**

Synopsis

Phoebe's father gives her a hot water bottle every birthday, every Christmas, and when she is 'extra specially good'. Phoebe hopes for a puppy, but still hot water bottles appear. When her father's shop catches fire Phoebe becomes a mobile fire brigade and uses her 158 hot water bottles to put out the fire. And what do you think she got for a present?

Focus

Counting, grouping, graphing.

Preparation

- Plain and coloured paper, graph paper and/ or centimetre grid paper
- A variety of concrete materials

Activities

★ Set the task by asking the children to represent 158 hot water bottles using a variety of materials. Younger children could do this by cutting out and pasting hot water bottle shapes. Older children may decide to cut up centimetre grid paper, or to represent the hot water bottles on graph paper. If the children are not familiar with graph work, the teacher might model this part, with contributions from the class.

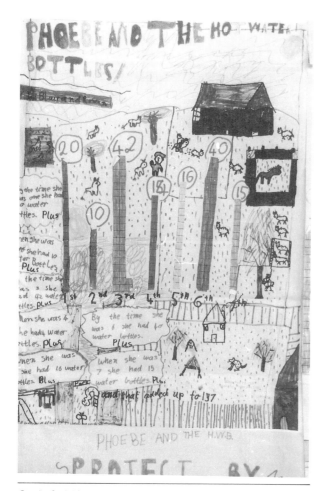

Grade 3 children at Wonga Park made a graph, wrote and drew about Phoebe's hot water bottles.

★ Ask the children to group the hot water bottles, choosing their own criteria. The book does not state the colour of the bottles, or how many Phoebe received each year, so the children can make their own decisions.

★ Ask the children to work in pairs or small groups to make a display of their groupings — graph, picture, concrete material etc. For children who have had limited experience in creating visual representations it may be appropriate to hold a class discussion and for the teacher to assist the children by modelling the process. Encourage the children to experiment with a variety of representations before deciding on the final form.

★ The children can then make up equations about their own and other children's displays.

★ Pose a problem for investigation: with one birthday and one Christmas per year, how many times was Phoebe 'extra specially good'?

Extension

★ Here is another problem: if she was good at the same rate how many hot water bottles would Phoebe have at age 9, 10, 12, 14?

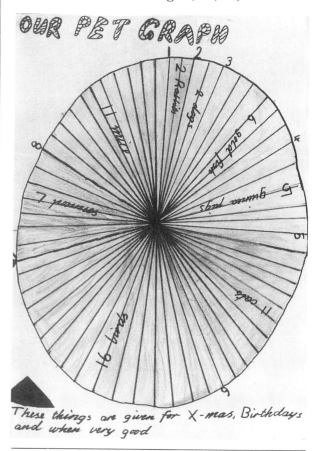

These things are given for X-mas, Birthdays and when very good

This graph of pets was based on the divisions of a clock face.

★ Ask the class how much water 158 hot water bottles hold. You will need one hot water bottle to start this investigation.

These boys at Waverley Park chose an unusual presentation for their graph of the number of soccer balls they received. How many did they get when they were 1, 2 and 3 years old?

★ The children could choose their own present for each time they are 'extra specially good', and make their own chart.

Indicators

● What sort of groupings did the children use to help them represent the hot water bottles?
● Did the children use place value to help them represent the hot water bottles?
● Did the children use symbols to represent more than one hot water bottle?
● Were the children's visual representations clear enough for another person to interpret?

Length of session

Two sessions may be needed to complete the main activity.

TEN APPLES UP ON TOP, Theo. LeSieg

Random House, 1988. **(M)**

Teacher information

The relation of the text to the illustrations is important in this activity. If only one copy of the book is available, some of the class might need to be involved in other activities (for instance acting out the story, or rewriting the text) whilst a small group is doing the main activity given below.

Synopsis

Three zany animals try to outdo each other balancing apples on their heads. After the final chase everyone has 'ten apples up on top'.

Focus

Counting, addition, subtraction, number comparison, equations.

Preparation

● Spike and loop abacus will be useful, in addition to counters.
● Multiple copies of the book, if possible.

Activities

★ Read the book to the children. Repeat the reading allowing the children to join in identifying the mathematics on each page.

★ The children can work in small groups and make up equations, number sentences or concrete representations of the numbers of apples on a page. They can then present their equation or other representation to another group, and ask them to identify the page which it matches.

★ The teacher may wish to direct children to particular operations. The book shows clearly subtraction as difference; children often have difficulty with this concept. (See, for instance, page 20: 'the tiger has seven, which is three more than the dog has'.)

Indicators

● What kinds of formal and informal recordings and representations do the children make?
● Note those who have difficulty with particular forms, such as equation writing.
● Are the children's equations or other representations accurate?
● Can the children interpret equations or number sentences accurately?
● Can they match them to an appropriate page?

Length of session

At least 60 minutes.

ALEXANDER, WHO USED TO BE RICH LAST SUNDAY,
Judith Viorst

Macmillan, 1980.   **(M)**

Synopsis

Alexander bemoans his poverty ... He had two dollars last Sunday and all he has left is a deck of cards with no seven of clubs or two of diamonds, a one-eyed bear, a half-melted candle, and ... toy money. The saga of how he spends or loses his money is told amusingly and realistically.

Focus

Money, writing.

Preparation

● Toy money, money stamps or real money.

Activities

★ Ask the children to write a story about how they might spend $2.00. They should check that the total spent (or lost) is exactly two dollars by using toy or real money, or drawings of coins (coin stamps are useful).

★ Find how much money each of Alexander's brothers had.

★ Check that Alexander's money adds up to exactly $2.00. Find out who got his money, and how much each person received.

Extension

★ Use catalogues to decide how to spend $50, or $100, or how to furnish your bedroom for $1000. This could be a small group activity.

I had two dollars last Sunday

Last Sunday it was pocket-money day and I got two dollars from my mum. My sister said I should buy something with it but I wanted to save it. First I went down the street to the shops. I used 25 cents on a packet of bubblegum but I dropped the packet on the way home so the next day I went to the shops again to get an icypole. But it melted on the way home—bye bye 40 cents. By that time I had wasted 65 cents. Now I just knew I was going to save the rest but my dad had to borrow 10 cents, he said he would pay me back but he forgot, I started to cry. So the next day I went to McDonalds, I ate lunch there which cost 1 dollar. I had 25 cents left, my mum told me to go to the shops to buy some milk with my 25 cents, when I came back I had nothing but toy money.

Grade 2/3 at Knoxfield wrote their own stories.

Plan a trip and decide how much money will be needed and how it will be spent.

> This was a very challenging exercise. Items costing $12·99, $6·75 caused problems! One pair suggested that we take amounts to the nearest dollar. All agreed this was the best solution.

Grade 2 children had to decide how to spend $100 on their families at Christmas using toy and outdoor-living catalogues.

Further reading

'Money moans', Roger McGough, in *Sky in the pie*
Alexander and the terrible horrible no-good very bad day, Judith Vorst (time)
See the section on **Money**, Chapter 8.

Indicators

- Does the money in the children's writing add up to $2.00?
- Has the child used realistic prices?
- What combinations of money has the child used?

Length of session

45–60 minutes.

DAD'S DIET,
Barbara Comber

Ashton Scholastic (Bookshelf), 1987. **(M, U)**

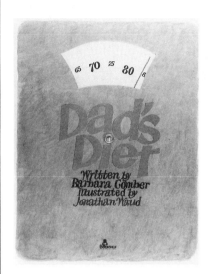

Teacher information

This activity allows children wide scope in choice of investigation.

Synopsis

Dad weighs himself and decides a diet is the only course of action. He is compared to the other members of the family at the beginning and end of the book, with pleasing results, despite obstacles.

Focus

Mass, fractions, ratio.

Preparation

- Bathroom scales
- Fraction kits.

Activities

★ Read the book to the children. Discuss their family experiences of diets and losing weight.

★ Ask the children to brainstorm as a class the mathematics they can identify in the book. Then ask them to devise problems for others based on the mathematics they have identified. They may come up with questions such as:

- How much weight did Dad lose?
- What are the weights of the children and of the mother?

- If Dad ends up weighing 6/7 of his previous weight, how many kilos has he lost?
- How long do you think it took him? (Page 20 may help.)

★ The children may also suggest measuring and weighing each other. They could graph their results. What is the average height and weight? They can measure and weigh other classes, graph these or find average heights and weights at different ages. This could lead into new investigations, such as when do children grow most quickly? Is it the same for height as for weight?

★ If sufficient data have been collected, perhaps from home as well as from school, the children may like to find another person whose weight is the same as theirs, ½ theirs, or 1½ theirs.

Extension

★ Look at the food eaten in the book. The children may like to investigate its nutritional value; and research to find out what constitutes a 'healthy' diet, and what would be a reasonable diet for losing weight safely.

★ Investigate body ratios — what is the ratio of wrist to neck, and of arm span to height?

Further reading

See the section on **Mass**, Chapter 8.

Indicators

- Can the children read the bathroom scales accurately?
- Can the children make and interpret a graph?
- Are the children able to operate with fractions (e.g. find out the weights of the children and the mother)?

Length of session

Depending which kinds of investigations are undertaken, this could take several sessions.

THE HILTON HEN HOUSE,
Jo Hinchcliffe

Ashton Scholastic, 1987.   **(M, U)**

Jo Hinchliffe — Illustrated by John Forrest

Synopsis

Farmer McMurray's house is over-crowded with a plethora of animals as well as his large family. He agrees to build a Hilton Hen House with a room for each of his ten hens, and ends by building a room for each member of the household. This is a counting book in which the numbers are not quite in order.

Focus

Counting, addition, spatial relations (plans).

Preparation

- Multiple copies, or a wall chart of the text, will be useful.

Activities

★ A good way for children to become familiar with the story is through dramatisation.

★ The teacher may pose, or the children may come up with, questions such as the following:
- How many different kinds of animals are there in the book?
- How many animals are there altogether?
- How many people are there?
- How many rooms does Farmer McMurray build?
- Which kinds of animals are missing from the picture on the second last page?

- Each person has a key; how many more will be needed if each animal has its own key too?

★ Draw a plan of the original house, marking the locations of the animals and people.

★ The children can draw plans and elevations for the new hen house. They should show all the rooms, marking who they are for. Make sure each animal has its room at the location indicated in the text.

★ Ask the children if all the rooms should be the same size. If not, which animals or people get the bigger rooms? Older children may indicate scale on their plans.

★ Draw side and back views of the house shown on the final page. Match the animals to the rooms in the picture.

Extension
★ Construct a Hilton Hen House.

Indicators
- Are the children systematic in their approach to the counting activities?
- Do the children match each animal/person to a room as indicated in the text?
- Are their plans clear to others?

Length of session
One session will be needed for the counting activities, but more if plans of the house are made.

A DAY ON THE AVENUE,
Robert Roennfeldt
Penguin, 1983.   (M, U)

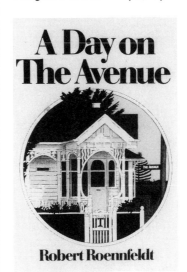

Synopsis
This is a no-text book showing a day in the life of a Geelong street. Insets show incidents such as two dogs knocking down a ladder and stranding a painter on the roof.

Focus
Time: day, drawing.

Activities
★ Show the book and discuss what is happening in the pictures. Ask the children to identify the theme of the book. Allow the children to study the book at leisure over a period of time. This book is full of detail for individuals and small groups to study, and multiple copies would be useful.

★ Ask the children to choose a place they are familiar with — their home, the classroom, the playground — and imagine what is happening there through the day. Ask the children to illustrate what is happening in their chosen place. Children can compare their illustrations and discuss similarities and differences.

Extension
★ The children can exchange drawings, and 'read' what is happening through the day.

★ They could write about their own or other children's drawings.

Further reading

See the section on **Time**, Chapter 8.
'Twenty-four hours', Charles Causley (in *Jack the treacle eater*)
Niki's walk, Jane Tanner (a no-text book for mapping)

Indicators

- Can the children sequence activities correctly through the day?
- Can the children identify what is likely to be happening at their chosen place through the day?
- Can they imagine what is happening when they are not there?
- Can the children follow each other's sequence of drawings?

Length of session

60 minutes.

A PET FOR MRS ARBUCKLE, Gwenda Smyth

Crown, 1984.   **(M, U)**

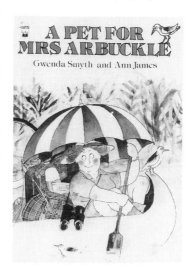

Synopsis

Mrs Arbuckle travels the world with a ginger-nut cat to seek the perfect pet, only to find it on her own doorstep.

Focus

Mapping, classification.

Preparation

- Outline maps of the world.
- Large world map for the class.
- Atlases and globe.
- Airline schedules, and schedules of airfares (the children could collect these).
- Calculators.

Activities

★ Read the book through to the children. On the second reading have the children mark on the world map the countries Mrs Arbuckle visits. Discussion of continents, capital cities and oceans could be developed.

★ On a blank map, in pairs or small groups, have the children mark in order the countries Mrs Arbuckle visited. They can calculate the distance travelled and the cost of the trip.

★ In their pairs or small groups, children can find the most efficient route she could have taken. They should consider the least distance, the least time and the least cost. Will each of these give the same result? A globe will be needed to find the shortest route between the countries.

★ Classify the animals in the story, naming their most important attributes, either as pets or zoo animals.

Extension

★ The children could investigate features such as life spans, size, diet and preferred climatic conditions of the animals in the story.

★ They can write an innovative text creating their own route and choosing other animals, in Australia or world wide.

Further reading

See the section on **Location**, Chapter 6.

Indicators

● Are the children able to identify countries, cities, oceans and continents on a world map, atlas or globe?

● What methods did the children use to find the distances travelled by Mrs Arbuckle?

● Were the children able to calculate the total distance travelled?

● How did the children decide on the most efficient route?

● Were they able to evaluate each other's decisions?

Length of session

Several sessons will be needed to complete the investigation satisfactorily.

SOMETHING ABSOLUTELY ENORMOUS,
Margaret Wild

Ashton Scholastic, 1984. **(M, U)**

Synopsis

Sally, with her enthusiasm for knitting, knits an enormous thing. The circus comes to town and the big top burns down. But Sally's thing is put to good use and saves the day. What does Sally do — go to the circus? No. We find Sally in the kitchen baking. The book ends with Sally about to bake something absolutely enormous.

Focus

Size, large numbers, creating and solving problems.

Activities

★ Brainstorm with the children the mathematics in the book. From this, the children may identify problems which they wish to work on.

★ List with the children vocabulary relating to size. In discussion, you may wish to focus on different kinds of measurement such as area, volume, height or length.

★ Discuss in groups what Sally could bake and how she could solve another problem. The children can draw and label what Sally baked, and write about how she solved the new problem.

> Sally could bake some giant doughnuts to use as life saving devices for the local beaches and swimming pools or:
>
> She could use them as wheels for bikes and cars. You can't get a flat with them.
>
> Children came up with these uses for giant doughnuts.

The VFL Pancake

Sally loved cooking so she decided to make something enormous. Sally cooked a giant pancake. While Sally was cooking the pancake the VFL park caught on fire because someone had dropped a cigarette on the ground and the grass caught fire. Sally kept adding more ingredients into the big pancake. At last she couldn't find anything that is meant to go in the pancake. Sally was going to wait for it to cool down but her mother said why don't we take it to the VFL park. I will get the neighbours to help us take it there so that Carlton and the Swans can play their game of football.

Children from grade 3 and 4 found new uses for Sally's baking.

★ Sally knits a thousand squares. Ask the children to make or represent a thousand: use Unifix, MAB, icy-pole sticks, graph paper, coloured paper etc.

Extension

★ Have the children count the squares on the circus top. Remember there were a thousand, so how many are round the other side of the tent? How big is a circus tent? Invite the children to research and find out.

★ Ask them: how long is a 50-gram ball of wool. Are all 50-gram balls of wool the same length?

★ What would they need to bake a cake big enough for the class? What about for the whole school?

Further reading

See the section on **Size**, Chapter 6.

Indicators

- Which areas of mathematics do the children focus on — e.g., pure number or applied number?
- Are the children able to identify a number of mathematical ideas in the book?
- Can the children list a range of vocabulary relating to size?
- Can they distinguish the different kinds of measurement — length, area, volume?
- Are they able to set up a problem?
- Do they really solve the problem which they set up?
- Were the children able to predict what a thousand would look like — e.g., how many 'one hundred grids' they would need?
- Could the children count their thousand accurately, and efficiently?

Length of session

This will depend on the teacher's and children's reactions to the book. It would take several sessions to complete the activities suggested.

THE TWELVE DAYS OF CHRISTMAS,
June Williams

Akers and Dorrington, 1984

THE TWELVE DAYS OF CHRISTMAS,
Brian Wildsmith

Oxford University Press, 1984

THE TWELVE DAYS OF CHRISTMAS,
Jack Kent

Hamish Hamilton, 1973.   **(M, U)**

Synopsis
These three books treat the traditional Christmas song in very different ways. June Williams' story has Australian animals persuading a reluctant swagman to rescue 'an emu up a gum tree'. Jack Kent treats the presents cumulatively — each day the narrator receives all the previous day's gifts together with the new ones for that day and is not always pleased!

Focus
Counting, addition.

Preparation
● Paper, scissors and paste will certainly be needed.
● Other materials needed will depend on whether the teacher decides to take an open-ended activity or a more structured approach.

Activities
★ If all three books are available, read them all to the class, inviting the children to join in the song. Compare the illustrations and discuss the approach of each illustrator, and, in the case of the Australian version, the author.

★ Using the Wildsmith or Williams version, ask the children to find out how many presents or animals were in the book.

★ With the Jack Kent book ask the children to find out how many items are on the last page. Some children may explore day four or five etc. This allows children to work at a task at their own level.

The Christmas tree made up of the gifts sent shows the triangular number pattern clearly.

★ The children may choose concrete material to count the presents or animals in the version being used. Ask the children to record and display their findings.

★ The children could measure squares and make a Christmas tree with one square for the first day, two for the second and so on. They could draw the presents on the squares. This activity is suitable for pairs.

★ The children may like to make a Christmas tree using cut-out hand shapes or handprints, and place the presents on the tree. These could be flaps with a puzzle on the front (e.g. how many presents on the fifth day?) and the answer and the drawing of the present underneath.

★ If the children have chosen their own method of recording ask them to report back to the class.

Extension
★ The children may choose to write class or individual books using their own presents or animals for the twelve days.

★ The children could classify and graph the presents.

★ Calendar work — ask them when is the twelfth day of Christmas.

★ Using Jack Kent's book, have the children investigate how many presents were sent over the whole twelve days.

Indicators

- What methods of recording did the children use?
- What materials or methods did the children use in finding the number of presents or animals?
- Did the children check their answers? How?

Length of session

Depending on the teacher's approach, this may take up to two sessions.

ALL IN A DAY,
Mitsumasa Anno

Putnam, 1990.   **(U)**

Teacher information

Peace is the theme of this book, and besides the mathematical activities suggested below, discussion of the theme is important. The book would make a good starting point for an integrated curriculum topic, as a stimulus for investigating life in other parts of the world and also for looking at the work of the contributing artists.

Synopsis

Anno and eight other well-known illustrators have contributed to this book. Each double page shows nine children around the world and what they are doing at the same time through twenty-four hours.

Focus

Time: seasons, clock.

Preparation

- A globe and atlases.

Activities

★ Show the book to the children and allow them time to browse through it and study the details.

Grade 3 and 4 children at Camberwell drew pictures of their day.

★ Discuss why the time, seasons and climate are different for the countries depicted. The globe and atlas will be useful here.

★ Ask the children to choose a country, locate it on the globe and in the atlas, and find out (at home or in the library) about its climate and customs.

★ In a small group, the children can choose a date and a time of day and find out what time and season it is in each of their chosen countries. As in *All in a day*, the children can illustrate and write about what is happening to a child in their chosen country at that time. Show the time and date for each picture.

★ Repeat this for a time (say) six hours later. Four pictures will then take each child through twenty-four hours.

★ Each group can make a book of their drawings.

Further reading

'Twenty-four hours', Charles Causley (in *Jack the treacle eater*)
See the section on **Time**, Chapter 8.

Indicators

● Can the children identify the seasons in another part of the world?
● Can the children explain the difference between seasons in the northern and southern hemispheres?
● Can the children work out the time in another part of the world, using the globe or atlas?
● Are the children able to identify the kinds of activities appropriate for their chosen time and country?

Length of session

At least two sessions of 60 minutes will be needed for the main activities. More will be needed for research on the countries chosen.

ANNO'S MYSTERIOUS MULTIPLYING JAR,
Mitsumasa and Masaichiro Anno
Putnam, 1983.   **(U)**

Synopsis

The mysterious jar contains a sea, in which is one island. On the island are two countries, in each country are three mountains, on each of which are four kingdoms. The pattern continues, beautifully illustrated, until we have eight cupboards each holding nine boxes. 'In each box there were ten jars. This story is about just one jar and what was inside it ...' The total number of each item is then shown, on successive pages, by red dots, one for the island, two for the countries, six for the mountains, up to a double page of 40 320 for the cupboards. The 362 880 boxes and 3 628 800 jars are not shown!

Focus

Multiplication, number patterns: factorials, large numbers.

Preparation

● Calculators (optional).

Activities

★ At the first reading, simply allow the children to enjoy the story. We suggest that at this stage you do not show the children the representations of the numbers in the second part of the book. On the second or a subsequent reading, estimate the number of items on each page. The children may compute on a calculator, or by hand. This is a good opportunity to practise rounding off.

> *Will it really take 180 pages to show the number of jars?*

★ Ask how many pages will be needed to show the number of boxes. And the number of jars?

> *Children really enjoyed the investigation of large numbers.*

★ Continue the story of the Christmas tree as suggested at the end of the book.

★ Make up new stories using this or another number pattern.

Extension

★ Investigate large numbers: e.g., how long is a million seconds? How many 1mm squares are there on a sheet of graph paper? How many grains of rice will cover one square metre?

★ Ask the children to write the largest number they can think of (give a time limit!) Now ask them to write a larger number.

★ Find large numbers in the newspaper or encyclopaedia.

★ Find a million of some item in the school.

Further reading

King Kaid of India, (in *The Victorian Fifth Reader*)

Millions and millions of cats, Wanda Gag

The phantom tollbooth, Norton Juster (chapter 15).

Indicators

- Can the children multiply accurately and use rounding off as they go through the book?
- Can the children predict how many pages will be needed to show the numbers of boxes and jars?
- Are the children able to use this pattern to create a new story?
- Are they able to think of and use another number pattern to create a story?

Length of session

At least one session, 45–60 minutes, will be needed for the main activity.

ROUND TRIP,
Ann Jonas

Scholastic, 1984.

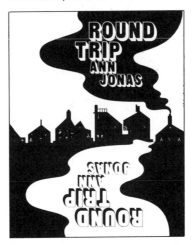

Teacher information

This book is read to the end and then turned upside down and read back to the beginning. The illustrations are turned upside down and a new text, continuing the narrative, is created. Because of the detailed illustrations, this book should be available to the children for individual examination before the activity is undertaken.

Synopsis

The black and white drawings illustrate a journey. When the end of the book is reached, turn it upside down and read back to the start again.

Focus

Spatial relations: shape

Preparation

A sheet of black and white cover paper, scissors and paste.

Activities

★ The book should be read to the children a number of times, with opportunities for discussion. The pages should be seen the 'right' way up and 'upside down'. Make the book available to the children for study and discussion.

★ A list of vocabulary can be drawn up which includes the variety of shapes and techniques used in the book: for example, the zig-zag road becoming lightning.

★ The children can draw simple shapes and turn them upside down to see what they could become. This process generates a lot of discussion about shape and about the environment.

These grade 4 children at Waverly Park captured the mood of *Round Trip* in their presentation of picture and text.

★ When the children have had the opportunity to become familiar with the process, have them create their own 'right way up' and 'upside down' pictures using the setting of day and night.

★ The children can then write a text to suit their pictures.

Indicators

● Can the children apply the technique to their own work?
● Do any of the children choose shapes beyond what is in the text?
● Did the children choose a variety of shapes?

KING KAID OF INDIA

The Victorian Readers: Fifth Book, Ministry of Education, Victoria (facsimile reprint), 1986.   (U)

Synopsis

This traditional story tells how the inventor of chess was invited to name his own reward. He asked for one grain of corn for the first square of the board, double that for the second square, double that again for the third, and so on. See page 2 for some background information.

Focus

Doubling, large numbers, volume, mass, money.

Preparation

● A chess board and pieces. Discussion of chess and other board games such as draughts before starting the activity would be an advantage.
● Grains of corn or rice. Rice is easy to obtain and to handle, so it may be more convenient to use this instead of corn in the activities.
● A variety of measuring equipment (scales, balances, litre measures etc.).
● Calculators are an option. The children will discover that these do not have the capacity for such large numbers.

Activities

★ Read or tell the story up to the stage where the king grants the wise man his reward, and asks his treasurer to calculate the number of grains which will be needed.

★ Ask the children to estimate the number of grains which will be needed, and to write up their estimates. This could be done in small groups, each of which could give a group estimate, or the children may choose to give individual estimates.

★ The children can then calculate the amount they think the wise man will receive. It may be necessary to work as a class over the first few squares. The teacher may prefer to limit the amount by looking initially at, say, the first thirty-two squares. Since calculators show only eight digits, the calculations will need to be done manually. Pairs or small groups are appropriate.

★ Finish reading the story. Compare the estimates made with the children's results and with the figure given in the story.

★ Discuss differences in calculations and also in methods of working out the large numbers.

★ Introduce other situations which include doubling — e.g. the puzzle of the waterlily which doubles in area each day. If it covers half the pond after ten days, when will it cover the whole pond? Or if the children offered to do the washing up for 1c the first day, 2c the next, 4c on the third day and so on, how much money would they be receiving at the end of the month?

★ Ask the children to present a report on their findings. This could be written, oral, a model or display, or a dramatisation of the story.

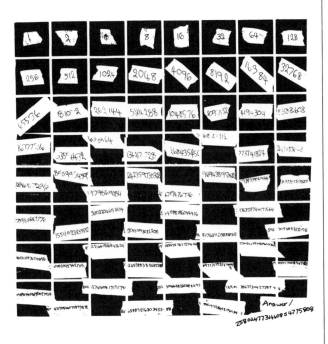

Grade 6 children worked in pairs to find the number of grains of corn on the chess board.

Extension

★ Discuss with the children investigations which they could undertake to do with the grains of corn in the story. They could discover the volume, mass or value of the corn, discover how many hectares of corn would be needed

to grow the crop (mark out a hectare in the schoolgrounds), and so on.

Further reading

Anno's mysterious multiplying jar, Mitsumasa Anno

The phantom tollbooth, Norton Juster (chapter 15).

Indicators

● How do the children cope with the large numbers involved?
● How far ahead can they predict in doubling?
● How did the children react when comparing their initial estimates with the results they got by calculating?
● What different methods did the children use when calculating?

Length of session

At least 60 minutes for the initial activity, and several sessions to discuss and extend the unit.

MR ARCHIMEDES' BATH,
Pamela Allen

Collins, 1985.   **(L, M, U)**

Teacher information

This book offers opportunities for work with children over a wide age range. However, the approach used by the teacher will vary, and some of the suggested activities are more suited for older children. Because the main activity involves water, teachers will need to decide whether to provide only one container of water for each group to use in turn, whilst other groups are involved in other activities, or whether to involve all the children in this messy activity at the same time. Going outside may solve the problem, as well as being very popular with the children.

Synopsis

Mr Archimedes becomes tired of cleaning up the mess when the bath continually overflows. Who is causing the mess? After measuring the depth of the water, he excludes the animals one by one from the bath, and discovers that it is himself and the animals who cause the mess. The fun begins.

Focus

Volume, water displacement.

Preparation

- Rulers.
- Containers with water.

- A variety of articles including zoo and farm animals and toy figures.

Activities

★ Allow the children to work in small groups and choose one toy figure and a few small animals. Ask the children to measure the water level and note the increases as toys are added to the water. The teacher may wish to model this activity, depending on the experience of the children.

★ Ask them how many animals it takes before the water reaches the top, overflows or reaches a certain mark.

★ After some experimentation, invite the children to choose some different toys or other objects to add, and estimate the increase in water depth. Check their accuracy by doing it.

★ Older children can make a graph of their findings, recording the articles used and the increase in water depth.

Extension

★ Children could write their own book using the animals they chose for the experiment.

★ Some children could be encouraged to write their findings in a formal report.

★ Who was Archimedes? Invite the children to research.

Further reading

Who sank the boat?, Pamela Allen
The crow and the pitcher, Aesop

Indicators

- Were the children able to predict what would happen to the water level?
- How did the children cope with accurate measurement?
- Did the children who wrote reports use appropriate language?

Length of session

At least 60–90 minutes. A second session may be needed.

WHO SANK THE BOAT?,
Pamela Allen

Putnam, 1990.   **(L, M, U)**

Teacher information

Because the main activity involves water, teachers may choose to provide one container of water for each small group to use in turn whilst other groups are involved in different activities. Alternatively, the activity could be done outside.

This book is available in big book format.

Synopsis

Five animal friends decide to go for a row on the bay. Presented in rhyme, the children are asked at each stage 'Do you know who sank the boat?' The mouse is finally blamed. All the animals return home dripping wet except the mouse.

Focus

Order, capacity, mass, balance, setting up and solving problems.

Preparation

- A variety of objects to be used as boats.
- Containers with water/a water tray
- Items for the children to make selections from to sink the boat. Ensure that material is different in size and mass, e.g. corks, plastic, nails etc. The children may like to collect these beforehand, if given notice.

Activities

★ Read the book to the children (if possible in big book format). Ask the children to discuss who sank the boat. The children may not realise that the mouse only 'sank the boat' because he was the last in, and that it is the sum of the animals' weights which in fact sinks the boat.

★ Draw the children's attention to the illustrations. Discuss the various positions of the boat and the change of water-line as the animals get into the boat.

★ Discuss how the text relates to where each animal sits and how their positions maintain the balance in the boat.

★ Allow the children to experiment with chosen objects in order to investigate which objects will change the water-line most, and what objects may cause the boat to capsize or sink. The children may like to make drawings to illustrate their findings.

★ Allow the children to work in small groups with a 'boat' and a number of objects. Ask the children to find five objects (to match the five animals in the story) which together will just sink the boat. Ask the children to decide and to record the order in which they are going to put them in the boat. Does the order matter?

★ Invite the children to repeat the activity using the same objects but in a different order. Give the children time to discover that the order has no relevance to the final outcome. Go back to the text. Did the mouse sink the boat?

★ The children could experiment with different numbers of objects needed to sink the boats. Ask what they think caused one boat to sink sooner than another. This could lead to a discussion on more/less mass, and what floats and sinks.

★ The children could report orally or in written or visual form to the class, and these reports could be used to investigate and clarify their understandings.

Extension

★ Teachers may like to extend into activities investigating balance and equality, using balance beams and scales and similar material to that used in the sinking the boat activity.

★ Rewrite the book using different animals or people, e.g. a group of pirates going out in a boat.

Further reading

Mr Archimedes' bath, Pamela Allen

Indicators

● Could the children show the water-line and the balance of the boat in their drawings?
● What conclusions about the sinking did the children reach?
● Were the children able to evaluate each other's reports and participate in discussion arising from them?
● Note children who need assistance with the floating/sinking concept.

Length of session

Depending on the investigations these activities may be done over three or four sessions.

ANNO'S COUNTING BOOK,
Mitsumasa Anno

Harper Junior Books, 1986.   **(L, M, U)**

Teacher information

Numbers are shown as cardinal, ordinal, time on a clock, time as months of the year, and as labels. The concepts of number shown are sophisticated, and the drawings appeal to all ages.

Synopsis

A textless picture book illustrating the numbers 0–12 through the growth of a township over months and years.

Focus

Number: cardinal and ordinal, counting, time.

Activities

★ Invite the children to comment on what they see in the book. If necessary, draw attention to the sequence of the seasons, changes over time etc. Give the children the opportunity to study and discuss the book in small groups.

★ The children may like to create their own counting book. Younger children will probably focus only on the cardinal aspect of number; older children may also include the sequence of the months, and could perhaps set the book over their own lifetime. This could be an individual or small-group activity.

Extension

★ *Anno's Counting Book* could be useful in any theme or topic which deals with time and change. See Donna Rawlins and Nadia Wheatley's *My place* for a child's perspective on a Sydney suburb over two-hundred years.

Further reading

See the section on **Counting books**, Chapter 7, and **Time**, Chapter 8.

Indicators

• Do the students identify the different uses of number in the book?

• Which uses of number do the children transfer into their own counting book?

• Did they use time in their book?

Length of session

A number of short sessions may be needed to create a book.

SIZES, Jan Pienkowski

Puffin, 1983.　　**(L, M, U)**

Teacher information

Although this book is usually regarded as a pre-school text, the concepts illustrated are quite subtle, and we have found that children up to at least grade 6 are challenged by the activity. Some upper grades may resist working with this text; it could be presented as a model for writing a similar book for younger children.

Synopsis

This concept book shows pairs of objects with similarities and differences — e.g., big lady, little boy; big whale, little fish; big block of flats, little house.

Focus

Size, classification.

Activities

★ As you read the book, ask the children what is the same and what is different about each pair. They usually find it easier to state the differences than the similarities.

★ Have the children draw their own pairs of objects — big and little, with a similarity and a difference. They can write what is the same and what is different about their objects.

★ Make a class book.

This activity can challenge a wide age range.

Further reading
See the section on **Size**, Chapter 6.

Indicators
- Are the children able to cope with the concepts of size, similarity and difference at the same time?
- Can the children explain the relationship between their objects?

Length of session
45–60 minutes.

5:

ACTIVITIES BASED ON RHYMES AND POEMS

My angles are many.
My sides are not few.
I'm the dodecahedron.
Who are you?

(*The phantom tollbooth*, p.146)

Verse is usually short, and therefore easily read as well as easy to remember. In this section, some rhymes and poems are used to develop mathematical activities. Besides the ones we have selected, there are many traditional rhymes which are suitable. A list of some useful anthologies is included in the bibliography.

HANDS
(*in* Ten little goblins)
and MEASURING
(*in* How many?),
Avelyn Davidson

Shortland, (Understanding Mathematics *series*), 1985.   (L)

Here are two rhymes about informal measuring.

Hands

One hand, two hands, three hands, four!
Four hands wide is the width of the door.
One, two, three, four, five, six, seven!
Seven hands tall is the height of Kevin.
Two hands, four hands, six hands, eight!
Eight hands long is the length of Kate.
Use your hands I'm sure you're able,
And find out the height, length, width
of a table.

Measuring

Today when I was playing
I found a piece of string.
It was just the right length
To use to measure things.
First I measured Daddy
He was ten strings tall.
Mummy was eight and a bit strings,
Then I measured the wall.
Four, five I counted
As I measured the window sill.
I wish I could measure our baby
But he won't keep still.
I'd like to measure an elephant,
or even a little mouse,
I wonder how many strings it would be
To the very top of our house?

Focus
Measuring length: informal units.

Preparation
● String.
● A variety of counters, sticks, straws, etc.

Activities
★ Read 'Hands' to the children.

★ Ask all the children to measure the length of their table using their hands.

★ Discuss how the children used their hands. Did they spread their fingers? Did they all get the same result? If not, why not? (Are all tables the same, are all children's handspans the same length, do all the children hold and place their hands the same way?)

★ The children may wish to continue using their hands to measure objects, or they may spontaneously suggest using other body parts (e.g. a foot, or one stride) for measuring. They will probably want to measure each other as in the rhyme.

★ Read 'Measuring' to the children.

★ Allow each pair of children to choose and cut a piece of string 'just the right length to use to measure things'. Ask them to find five objects to measure with their piece of string (e.g. each other, the door, table or blackboard).

★ When the children report back to the class, draw attention to two groups who have chosen to measure the same object. Consider why their results are not the same. Or are they the same?

★ Allow the children to experiment measuring a variety of objects with a variety of materials over a period of time.

★ Ask the children to estimate, and then find out, distances such as:
● How many sticks across the table?
● How many straws up the door?
● How many footsteps across the basketball court?
● How many bodies across the room?

★ This kind of investigation leads naturally into discussion of the need for a standard ('agreed') unit for measuring length. How are things such as material and tape measured when we buy them?

Further reading
See the section on **Size**, Chapter 6.

Indicators
● Do the children line up their units carefully when measuring?
● Do the children understand that the number of objects will depend on the size of the unit chosen?
● Do they understand that the smaller the unit, the greater the number of objects used?
● Can the children make reasonable estimates of short and/or long distances?
● Do the children use the language of length clearly and accurately?

Length of session
A number of short sessions will probably be most beneficial, over a period of time.

AUTUMN and APPLES,
Avelyn Davidson

in Skittles, Shortland (Understanding Mathematics series), 1985.   (L, M)

These two rhymes dealing with trees explore subtraction through the falling of leaves and fruit.

Autumn

Leaves were growing on a tree
The wind has blown some down I see
Four have drifted far away,
Sixteen upon the branches stay.

Leaves were growing on a tree,
The wind has blown more down I see.
Eight have drifted far away,
Twelve upon the branches stay.

Leaves were growing on a tree,
The wind has blown more down I see.
Fifteen have drifted far away,
Five upon the branches stay.

Leaves were growing on a tree,
Five more have blown down I see
Twenty leaves, the tree is bare,
Autumn's gone and winter's here.

Apples

I found a lovely apple tree,
Lots of apples are for me.
I shook the tree as hard as I could,
Down came some apples, mmm they're good.
How many apples were on that tree?
How many apples fell on me?
Seventeen apples, minus four,
I think I'll shake it just once more.
Thirteen apples on the tree,
This time eight apples fell on me.
How many left upon the tree?
Count the apples and you'll see.
The first time that I shook the tree,
Four rosy apples fell on me.
The second time I shook off eight,
So that makes twelve minus one I ate.

Focus
Subtraction, equations.

Activities
★ Read both rhymes to the children. This book is available in large format, which would be ideal. Ask the children what is happening in the rhymes.

★ Invite the children to work in groups and to record the equations in the rhymes in their preferred way. Have the groups report back and show their recordings.

★ Ask the children to make either the autumn or the apple tree, and demonstrate the rhyme. These trees could become a way of practising subtraction, by allowing the children to make up new stories and therefore new equations.

★ Children could make trees for other fruits, and possibly adapt to addition. The teacher could model this by making a new rhyme such as:

I found a lovely blossom tree
Lots of apples growing for me
Growing fast, what a treat,
Lots of apples for me to eat.
Seven apples, a week's lunch
Red and smooth hear them crunch.

Now there's enough to make a cake,
I see there are another eight.
Ten more apples for a pie
Another piece, we all cry!
Four apples for us to share
Soon the tree will be quite bare!

Imagine what could be done with a grapevine!

Extension
★ Explore other counting and number rhymes and allow the children to explore ways of changing or innovating on both equation and presentation.

Further reading
Mr Brown's magnificent apple tree, Y. Winer

Indicators

- What informal and formal mathematical language did the children use?
- What informal and formal recording methods did the children use?
- Were there any children who may need extra time exploring subtraction?

Length of session

Depending on the extent of exploration of equations and innovative texts, these rhymes supported by others could be used over many sessions.

A DREADFUL THOUGHT,
Rose Fyleman

in Wheels, Shortland, (Understanding Mathematics series), 1985.

A SPIDER'S BEDSOCKS,
Phil Mena

in The big bed, *Department of Education, New Zealand, 1984.*   **(L, M)**

Here are two poems about knitting socks and stockings for spiders.

A dreadful thought

Oh! Mrs Spider,
How dreadful it would be,
If your children needed stockings,
Like Jennifer and me!

For every little spider,
Never mind how small,
Every little spider
Has eight legs in all!

Four pairs of stockings,
Four for every one,
You never, never, never,
Would get your knitting done!

A spider's bedsocks

A spider knitted bedsocks —
He made them big and bright.
He put them on his little feet
To keep them warm at night.
And when they needed washing
He used a lot of pegs
To hang the socks outside to dry —
Eight pegs, eight socks, eight legs.

Focus

Multiplication, 1:1 correspondence.

Preparation

- Make charts of the rhymes so that the children can see and read the text. 'A dreadful thought' is available as a big book.

Activities

★ Read the rhymes to the children.

★ Identify how many legs a spider has. Using 'A spider's bedsocks', ask the children to draw the spider's body, and add the correct number of legs. Ask them to draw a washing line and put on it the correct number of pegs and socks for their spider.

★ Some children may like to draw more spiders — but they must give each one eight legs and the correct number of pegs and socks on its washing line. A class or group mural may be made, and the children can then find out the total number of legs, pegs or socks. Ask the children whether they think there will be the same number of each. Young children able to match objects may not have abstracted the concept of equal numbers.

★ Ask the children to find out how many stockings Mrs Spider will have to knit. They can use the picture to find out how many children she has, the teacher may give a number, or the children may choose a number. If appropriate the teacher may like to start this activity by modelling with a particular number of baby spiders, and then let the children choose their own number.

★ The children could then identify how many pairs of stockings Mrs Spider has to knit.

Extension
★ Ask the children to make up problems about ants, centipedes, millipedes etc.

Further reading
Goldilocks and the three bears, page 20.

Indicators
● Can the children match pegs, washing and spider legs?
● How did the children approach the problem of counting Mrs Spider's stockings? Note which children counted by ones, and which grouped the stockings (e.g. by twos, fours or eights).

Length of session
45–60 minutes.

MICE,
Rose Fyleman
in Hilda Boswell's, Treasury of Poetry, Collins   **(L, M)**

I think mice
  Are rather nice
Their tails are long,
  Their faces small,
They haven't any
  Chins at all.
Their ears are pink,
  Their teeth are white,
They run about
  The house at night.
They nibble things
  They shouldn't touch
And no-one seems
  To like them much.
But I think mice
  Are nice!

Focus
Modelling to solve a problem, ordering information.

Activities
★ Discuss the different parts of the mice mentioned in the poem and other parts mice need (eyes, legs etc.). With the children, establish a short list of the most important parts (three or four) needed to make a mouse. These could be eyes, ears, legs and tail.

★ Now use the parts decided on to make a mouse. Draw on the blackboard, cut and paste on a large sheet of paper or card, or model using blocks, straws, counters etc. This will establish that a mouse has, for instance, one tail, two eyes and four legs.

★ Now pose the problem (changed if necessary to fit the parts chosen):
  I have 9 tails,
       8 eyes,
      10 ears,
      15 legs.
  How many complete mice can I make?
(Do not make the numbers too high; our classroom experience has shown that this can be

a difficult problem for 6–8-year-olds even with the numbers given here.)

> Mike and Patrick worked very systematically. 8 = 4 pairs eyes, 10 = 5 pairs ears, 9 = 9 tails.

> Some children took out groups of counters and didn't know what to do next.

> These girls were convinced that there were enough legs for 4 mice. I asked how many legs have you drawn? 16. How many legs did you have? 15. Did you have enough? Yes.

★ Allow the children to draw, write or use concrete material to solve the problem. Ask them to explain and justify their solutions. Discuss the methods of solution as a class.

Extension
★ Ask the children to make up their own problems using this model. For example, they could think about the parts needed for a billy cart or a bicycle.

Indicators
● Do the children use all the information given, or do they assume there will be nine mice because there are nine tails?
● Which children tackle the problem systematically?

Length of session
30–45 minutes.

BAND-AIDS,
Shel Silverstein

in Where the sidewalk ends, *Harper Junior Books, 1984.*
(L, M)

> I have a Band-Aid on my finger,
> One on my knee, and one on my nose,
> One on my heel, and two on my shoulder,
> Three on my elbow, and nine on my toes.
> Two on my wrist, and one on my ankle,
> One on my chin, and one on my thigh,
> Four on my belly, and five on my bottom,
> One on my forehead, and one on my eye.
> One on my neck, and in case I might
>   need 'em
> I have a box full of thirty-five more.
> But oh! I do think it's sort of a pity
> I don't have a cut or a sore!

Focus
Counting, addition, subtraction, re-grouping.

Preparation
● Have a variety of concrete materials available, such as Unifix, straws, Plasticine, MAB, icy-pole sticks, counters, an abacus, a bead frame.
● You will need a large sheet of paper and Band-Aids for the extension activities.

Activities
★ Let the children estimate how many Band-Aids the child in the poem has used. Make a list of these and keep it till the end of the session.

> When children reported back there was lots of discussion on the best way to count — by ones, twos, fives or tens.

★ Ask the children to work in pairs to find the number of Band-Aids on the child in the poem, and also the total number of Band-Aids in the poem.

★ Ask the children to check their answers by using different materials. The children can

Mrs Griffiths read us a poem named 'Band-Aids'. When she finished she asked us how many Band-Aids were on Peter. How Erin and I worked it out, well I sat on the mat while Erin went to get some paper. I copied from the book. The number of Band-Aids on Peter was 26. Erin and I thought there was 22 Band-Aids on Peter but I was wrong.

All of 1N and 1S had to work out two different problems. One problem was to work out how many Band-Aids Peter had on himself.
   And the second problem was to add up the other Band-Aids with Peter's Band-Aids.
   We also used straws to help us work it out.

Grade 1 children at The Basin wrote about solving the problems from Avelyn Davidson's poem on 'Band-Aids'.

● Could the children check their work?
● In reporting, could the children justify their methods?
● Could they articulate their findings clearly?

Length of session
60–90 minutes.

then report to the class how they approached the task and how they used the materials. Ask which materials they found most helpful, and why. Re-grouping may be discussed here.

★ Suggest that the children compare their estimates with their recordings.

Extension
★ Have a small group work with a life-size tracing of a child, and place the Band-Aids according to the poem. Group the 'thirty-five left' underneath. How many out of a packet of 100 Band-Aids have been used?

Further reading
'Band-Aids' , by Avelyn Davidson in *Wheels*, Shortland (*Understanding Mathematics* series), 1984 (more suitable for younger children).

Indicators
● Which material did the children choose? Why?
● Did the children work through the poem systematically?
● Did the children re-group or count by ones? (This information could be used to set up further experiences in re-grouping.)
● What formal and informal mathematical language did the children use?

AGE,
Avelyn Davidson

in *Skittles, Shortland,* (Understanding Mathematics series), 1985   **(M, U)**

> Mary's five and Bea is three,
> Bob is nine, that's three times Bea.
> In four more years I'll be eleven,
> That's much better than being seven.
> Grandad's sixty I was told,
> How many years till I'm that old?
> Funny how Mum's age never changes,
> She's been twenty-one for ages.

Focus

Computation: the four operations, creating equations and problems.

Activities

★ Read the big-book version to the class. Focus on the mathematical relations in the poem — how many times older than Bea is Bob? How old is the person talking?

★ Have the children calculate how many years till the person talking is as old as Grandad. Ask the children to explain how they worked out their answers. Ask them to identify a part of the poem which must be untrue, and why they think it is untrue.

★ Ask the children to write down the children in the poem in order of age. Some will have difficulty with the 'I' of the poem — who the 'I' is, what his or her name is, and how old he or she is.

★ Ask the children to write down the names and ages of their own family members.

> I didn't know how old my mum was but I worked it out from knowing she was born in 1951.

★ Now ask the children to make up a problem about the ages of the people in their family.

Give some examples, preferably from your own family, to get them started. Give a simple one-step problem, and perhaps a two-step problem, and also one with insufficient information. For example:

- Deryn is twenty-one, and she is two years older than Morwenna. How old is Morwenna?
- Deryn's father is twice as old as she is, and six more years. How old is he?
- Tim's mother is twenty-eight years older than Tim. How old is she?

Discuss these problems, especially the last one.

I am 8 my dad is $(4 \times 8) + (2 \times 3)$. My brother is 5×2 and my mum is 12×3.

Elisa is 15, Dad is 37, who's the oldest?

I am 9 and my Dad's 40. How many more years until I am that old?

I am nine, my brother is nine, my mum is twenty-two years older than me. How old is she?

My mum is 35 and my grandfather double that age.

Samples of problems made up by grade 3 children at The Basin.

★ The children can give their problems to each other to solve.

★ Compile a book of these problems. The children could work on this task in co-operative groups of three or four.

Indicators

- Do the children identify that Mum cannot really be twenty-one?
- Are the children able to make up one-step problems, and two-step problems?
- Which processes or operations do they use in their problems?
- Are they able to give sufficient information in their problems?
- What difficulties do they have in solving the teacher-directed problems and each other's problems?

Length of session

60–90 minutes.

THE SONG OF THE SHAPES,
Charles Causley

in Jack the treacle eater, *Macmillan, 1987 (**U**)*

Miss Triangle, Miss Rectangle
   Miss Circle and Miss Square
Went walking down on Shipshape Shore
   To taste the sea-salt air.
They talked of this, they talked of that,
   From a, b, c, to z.
But most of all they talked about
   The day they would be wed.

'My sweetheart is a sailor blue!'
   Miss Circle sang with joy,
'And dreams of me when out at sea
   Or swinging round the buoy;
When homeward bound for Plymouth
Sound
   Or sailing by the Nore,
Or gazing through the port-hole
   At some shining foreign shore.'

'And when we both are married,
   Up at the Villa Sphere
We'll have a Christmas pudding
   Each evening of the year;
And ring-a-ring o'roses
   We'll dance at night and noon,
Whether the sun is shining
   Or if it is the moon.'

'Good gracious!' cried Miss Triangle,
   'But it's quite plain to see
What might be best for you and yours
   Won't do for mine and me.
He plays the balalaika
   In a Russian Gypsy Band
All up and down the country
   And in many a foreign land.

He loves the dusty desert,
   The camels and the sun,
And sits and thinks beside the Sphinx
   Of when we shall be one.
He's bought for us a dream-house
   By shady palm-trees hid.
Do say you all will pay a call
   At Little Pyramid.'

'My boy so rare,' then said Miss Square,
   'Is different yet again.
His world is one of timber,
   Of chisel and of plane.
I might have wed a constable,
   I might have wed a vet,
But fairly on a carpenter
   My heart is squarely set.

Our own dear home won't be a cone,
   A cylinder (or tube),
But just a quiet cottage
   At the village of All Cube.
And you must come and visit us
   Each Saturday at three
For sticky buns and cake (with plums)
   And sugar-lumps and tea.'

'Though pleasant,' sighed Miss Rectangle,
   'To walk among these rocks is,
I'd sooner far be where my dar-
   ling's making cardboard boxes.
He makes them big, he makes them small,
   He makes them short and long,
And all the day (his workmates say)
   He sings a sort of song.

'My bride and I one day will fly
   Beside the Spanish sea
And live in Casa Cuboid
   Which we've built above the quay.
It has a special corner
   For fishing with a line
And catching of fish-fingers
   For friends who come to dine.'

Miss Triangle, Miss Rectangle,
   Miss Circle and Miss Square
Came walking up from Shipshape Shore
   Without a single care.
But pray remember if you play
   The match and marriage game
Opposites often suit as well
   As those who seem the same.

Focus
Shape, writing.

Preparation

- A set of three-dimensional geometric solids as well as pattern blocks and other two-dimensional shapes would be useful.

Activities

★ After reading the poem, discuss the shapes mentioned. Identify the distinguishing features of each.

★ Ask the children to identify with a chosen shape (one from the poem or another). In small groups, they can list other geometric shapes which include their chosen shape, and everyday objects which incorporate their chosen shape. They may find it helpful also to list words which describe their shape or its features — for instance, for a cone they might list words such as round bottom, circular base, rolls, pointy top, flat face, curved surface and so on.

★ Now ask them to write about their chosen shape, their hobbies, jobs, house and so on as in the poem. Remind them of the final verse, where Causley suggests that opposites can be as successful as 'those that seem the same'. Do not insist — or even suggest — that they write in verse as this will inhibit their ideas.

★ Share the writing. Some children may like to make a book.

Extension

★ The children could make from paper or card the three-dimensional shapes they have used in their writing.

Indicators

- Do the children know the names and properties of shapes? (Note children who have difficulty, and shapes which are not well known.)
- Can the children describe shapes clearly and accurately? Note their use of formal and informal language.
- Can the children identify a variety of objects which incorporate their chosen shape?

Length of sesssion

60 minutes.

6:

THEMES

'To tall men I'm a midget, and to short men I'm a giant; to the skinny ones I'm a fat man, and to the fat ones I'm a thin man. That way I can hold four jobs at once.'

(The phantom tollbooth, p. 96)

INTRODUCTION

In this section we look at groups of books which could be used in thematic work. We have used two kinds of themes:

- General (e.g. bed-time)
- Mathematical (e.g. size)

In each case, we discuss the kinds of mathematical work which might arise from the suggested books. Whilst not suggesting that other curriculum areas be neglected, we regard many curriculum areas as naturally integrated with the mathematics: art, language, social studies, science. Teachers will naturally develop other activites which will extend children's thinking across the curriculum.

We have not developed lesson plans in this section; rather we present some ideas for teachers and children to develop. It is expected that they will add many more as they explore the books and themes together.

In developing a theme, both open-ended and teacher-directed activities may be used. One profitable approach is to invite children to identify the mathematics in the books or in the topic. Such an approach can set the direction of the theme and allows children to devise their own investigations and explorations.

PREPARATION

Collect the books to be used. Multiple copies may be useful. Read them to the children, or allow children to read them and browse through them. It is important for the children to be familiar with the books before they start activities.

Have available a wide selection of materials:

- A variety of paper and card (coloured, grid, writing...).
- Scissors, paste, a stapler etc.
- concrete materials (both structured and unstructured).
- A variety of measuring equipment.
- Calculators.

APPROACHES

Either a teacher-directed or a child-centred approach, or some combination, can be used. In either case, the teacher should read the books to the class and to small groups, and allow the children to read, browse and explore the books before starting activities. If using a child-centred approach, then brainstorm as a class the mathematics, and (if wanted) other curriculum areas, in the books.

Invite the children to choose an area and to devise an investigation. Small groups could work together on a chosen task. If using a teacher-directed approach, the teacher will set up the explorations or investigations. The whole class may be working on the same topic, or small groups may be given or choose alternative topics.

EVALUATION

The children's progress may be monitored using the indicators suggested in the class lessons as appropriate.

THEME: BED-TIME

Introduction

These books are all to do with bed-time, or, in the case of Alex's bed, the bedroom. There is much ritual associated with this time of day and it can be used to show how sequence and order are reflected in our routine day-to-day living. When is the routine broken? Why?

Booklist

Ten nine eight, Molly Bang. (**L**): A countdown to bed-time, beginning with '10 small toes all washed and warm' and ending with '1 big girl all ready for bed'.

Ten sleepy sheep, Holly Keller. (**L**): What you do when you can't sleep and decide to count sheep.

One dragon's dream, Peter Pavey. (**M,U**): About what might happen when you are asleep. The rich illustrations will be pored over by children of all ages.

Winifred's new bed, L. and R. Howell. (**L**): Winifred finally grows out of her cot but finds her new bed too big. Each night she invites a toy in until the bed is full.

Alex's bed, Mary Dickinson. (**M,U**): This book has a different focus. Alex's room is a mess. After a lot of thought, he and his mum decide to build a high bed in the 'wasted space'.

The quilt, Ann Jonas. (**L,M,U**): During the night, the patchwork quilt turns into a landscape.

Activities

Time:

★ List and order pre-bed-time activities (*Ten nine eight*). The children can take home a sheet to record these, and note the times for each. Do they do things in the same order every night? Do they do them at the same time?

★ Individual or class recordings can be made, using clock faces, narrative, time-lines or other methods. Is there a night in the week when the routine changes? Why does it change?

Counting:

★ Counting forwards, backwards, cardinal and ordinal arise from the counting books and from the toys in *Winifred's new bed*.

★ The children can compare their own toys with Winifred's, and make their own list of toys for a week. How many would they need for a month or a year? How many can they fit in their own bed?

Classification:

★ The children can list and group the objects on a page of *One dragon's dream*. They can compare and discuss their lists and grouping criteria.

Spatial relations:

★ The endpapers in *Winifred's new bed* show a regular pattern of her toys. The children can look for symmetry, find the repeats and draw lines (on a photocopy or tracing paper) to join the same toys. What patterns do these lines make? The children can draw their own toy patterns. (This could make lovely wrapping paper or greeting cards.)

★ Using *The quilt* the children can design and make their own patchwork quilts. Materials could be coloured paper, wrapping paper or fabric scraps. After free experimentation, the teacher may like to introduce constraints such as 'no two the same next to each other', or 'only four patterns', or 'no two the same in a row'.

★ Using the idea in the book, the children could create their own journey through their quilt. This could involve mapping skills and the writing or telling of a narrative.

★ *Alex's bed* could lead to a discussion of space. The children could draw plans of their own bedrooms, and discuss Alex's problem in terms of their own rooms. Why didn't the use of Alex's 'wasted space' solve his problem?

★ What do you need in a bedroom? Where does everything best fit? Have the children design their ideal bedroom. Catalogues could be used to help cost the furnishings and decorating.

★ Role play the discussion between Alex and his mother on how to make the bed. Use wood offcuts or other material (plastic Meccano, icypole sticks etc.) to build the bed.

THEME: SIZE
Introduction
This topic is extensive, and covers many concepts, including ordering, comparison, measuring (both formal and informal), length, area and volume. To develop such concepts, much oral work as well as free exploration and experimentation should precede and complement more formal activities.

In our treatment we have listed the books and rhymes used under the headings of:
● Ordering
● Measuring
● Over-size
● Under-size

Activities have been listed in the same order, although there is some overlap as one activity often leads on to another with a different emphasis. We are not suggesting that these be undertaken at one time, or worked through systematically. Rather, these activities should be used to enhance and enrich your mathematics lessons when appropriate within your classroom program. Because many activities overlap, a checklist to monitor the concepts and skills covered would be useful. Such recordings would contribute to evaluation of the classroom program and to assessment of the children.

Many of the suggested activities are stimuli to investigations. When the children are given the opportunity to contribute, they come up with many ideas which will extend the activities outlined here.

Non-fiction material should of course be used with the fiction and the concept books which we have listed. Some particularly useful resources are mentioned in the text; many others will be found in school libraries and at home.

Booklist
Ordering:
Dear zoo, Rod Campbell. (**L**): Finding a pet of the right size and temperament can be difficult.
The bad-tempered ladybird, Eric Carle. (**L,M**): The ladybird spends the day looking for an animal big enough to fight.

Titch, Pat Hutchins. (**L,M**): Titch is the smallest of three children, and has the smallest of everything — until he plants a seed.

You'll soon grow into them, Titch, Pat Hutchins. (**L,M**): Titch has his brother's and sister's cast-offs, but eventually is able to hand them on in turn.

Happy birthday Sam, Pat Hutchins. (**L,M**): On his birthday Sam is still not tall enough to reach the door handles, light switch or taps — but Grandpa's present solves his problem. See page 27.

Sizes, Jan Pienkowski. (**L,M,U**): A concept book which focuses on size, difference and similarity. See page 53.

All fall down, Brian Wildsmith. (**L**): A balancing act with animals in order of size.

The cat on the mat, Brian Wildsmith. (**L**): The cat is followed by animals in increasing order of size until there's no room left on the mat.

Frank (grade 5/6) has used the concepts of size, similarity and difference in his drawing.

Rhymes and poems

'Six little firemen' (in *What's the time Mr Wolf?*), Avelyn Davidson. (**L**): A rhyme about six firemen and their sizes in relation to each other.

'What size are you?' (in *How big is big?*), Avelyn Davidson. (**L**).

'What is big?' (in *Sounds of numbers*), Henry Ritchett Wing. (**L,M**): A rhyme comparing Tommy to a sequence of larger and a sequence of smaller animals.

'The end' (in <u>*Now We Are Six*</u>), A. A. Milne. (**L, M**): A rhyme detailing the advantage of being six over any previous age.

Measuring:

How far can you jump?, Keith Pigdon and Marilyn Woolley. (**L,M**): A comparison of the jumping prowess of some animals.

River Red, Keith Pigdon and Marilyn Woolley. (**L,M,U**): The growth of a river red gum is described and compared to familiar objects.

'Hands' (in *Ten little goblins*), Avelyn Davidson. (**L**): A rhyme to stimulate informal measuring. See page 55.

'Measuring' (in *How many?*), Avelyn Davidson. (**L**): A rhyme about measuring with a piece of string. See page 55.

Over-size:

Jim and the beanstalk, Raymond Briggs. (**M,U**): Jim provides the giant with giant spectacles, false teeth and a wig — and escapes.

The biggest cake in the world, Joy Cowley. (**L**): What ingredients will be needed for the biggest cake in the world?

The BFG, Roald Dahl. (**L,M**): The Big Friendly Giant takes Sophie on a dream collecting expedition.

Amanda and the magic garden, J. Himmelman. (**L,M**): Not only do the magic seeds grow giant vegetables, but the animals who eat the vegetables grow too.

Mrs Bubble's baby, Margaret Mahy. (**L,M**): The cure for Mrs Bubble's tiny baby sends him through the roof.

Gulliver's travels, Jonathan Swift. (**M,U**): This classic includes Gulliver's visit to Brobdingnag, the land of the giants.

Something absolutely enormous, Margaret Wild. (**M,U**): Sally knits something absolutely enormous. How can it be used? See page 41.

Jack and the beanstalk. (**L**): The well-known tale of Jack's climb up the magic beanstalk and visit to the giant.

The Derby ram, William Stobbs. (**L**): A traditional rhyme about a gigantic ram.

Under-size:

Arabella, the smallest girl in the world, Mem Fox. (**L,M**): Tiny Arabella has everything just like yours or mine, but uses her belongings in an individual way.

The shrinking of Treehorn, Florence Heide. (**M,U**): Treehorn shrinks, but nobody believes him.

The borrowers, Mary Norton. (**M,U**): The borrowers live under the floorboards, and 'borrow' all those missing items — pins, pencils, cotton reels etc. — to use in their own way.

Gulliver's travels, Jonathan Swift. (**M,U**): The most famous of Gulliver's adventures is his visit to Lilliput.

Thumbelina, Hans Christian Andersen. (**M,U**): The well-known story of the girl the size of your thumb.

Activities

Ordering:

Introductory activities

We recommend that the following oral and movement activities be done before reading the books and before starting on more formal activities, as well as being interspersed in the sessions as appropriate.

★ Invite the children to order themselves by size. Make a chart, graph or frieze. This could be done at intervals through the year.

★ Ask the children, in groups, to collect a number of items and put them in order. The items could be selected randomly, or could be similar — for instance, cut straws.

★ Give each group of children an item as starting point (or allow them to select their own), and ask the group to find five items getting bigger, or five getting smaller.

★ Ask the children to describe what they have done in their ordering activity, using relevant vocabulary such as big, bigger, biggest; small, middle, large; thick, thin; tall, short and so on. Listening to the children and noting the vocabulary that they use correctly and comfortably will help teachers to gauge their level of understanding. It is important for teachers to model correct language whilst accepting children's approximations towards correct usage. After oral work the teacher and children can make a chart of vocabulary, looking particularly at comparisons and opposites.

Book activities

★ After reading and discussing the order of size in *Goldilocks and the three bears* (see page 20), the children may like to bring their own toy or teddy-bear. This could be the stimulus

for a whole host of measuring activities. One interesting ordering activity is to ask children to join in groups of three, with a big, middle-sized and small toy in each group. Ask the children to put their toys in order, and to draw the three toys in order, either as a group or individually.

★ After reading the rhyme 'Six little firemen', ask the children to draw the firemen in order of size. Note difficulties individual children have in drawing, ordering or labelling. The children may also like to act out the rhyme. They would have to choose a child for the first fireman and work from there.

★ *Titch* and *You'll soon grow into them, Titch* are good starting points for discussion about sizes of family members. The children can draw their own family members in order of size. Some children may like to include their pets. The text of *Titch* could be re-written with a new twist at the end (e.g. the knitting that grew and grew and grew, or the Playdoh that made a long snake that grew and grew and grew).

★ As you read 'What is big?' to the children, ask them to find and order objects to represent Tommy and the animals in the rhyme.

★ Both *Dear zoo* and *The bad-tempered ladybird* show animals in order of size. The children can re-write these texts using animals of their own choice. These books can also be used to initiate investigation into the sizes of animals. Research in the library or at home will be needed. The *Informazing* and *Prehistoric giants* series (Methuen) are useful references both for information and for presentation. The results could be graphed, or sizes of animals shown on a grid.

Measuring:

The activities listed under ordering can lead naturally into measuring, both formal and informal. The rhymes 'Measuring' and 'Hands' will provide a stimulus for undertaking similar measuring activities, using a variety of material for informal units and measuring a variety of objects. The children can be asked to repeat some of the activities at home, with the help of their family. Many such experiences are needed to develop and consolidate the concepts and skills involved in measuring. The children can work in small groups, and should compare and discuss their results. Estimating and the comparison of their estimates with their measuring are important, because children develop a feel for measurement and confidence in their measuring ability through such activities.

★ After reading *Happy birthday Sam* ask the children to first estimate and then find out how far up the wall they can reach. Ask them to estimate how high up different children in the class and the teacher can reach. If they stood on a chair, how high could they reach? Experiment and record the findings, compare with others. Who stood on tiptoe, who on flat feet? Use string or streamers or formal units as appropriate for their stage of development.

★ Using the same book and the poem 'The end' from <u>Now We Are Six</u>, ask the children to think back to a time when they couldn't reach the door handle, or the sink or some other object. How old do they think they were then? Investigate size at different ages. They could undertake this practically by measuring little brothers or sisters at home and older children at school. Class or individual books could be made, showing children's sizes and the things they could reach at different ages.

★ Upper grades could graph ages and heights, either averaging the data they have collected or indicating ranges of heights at different ages. At what stages is growth most rapid? Does an age/weight graph show the same pattern of growth?

★ Ask the children to investigate their local supermarket. What items are at eye level for three- to five-year-olds? Why might these items be placed at that level?

★ Follow *How far can you jump?* by an athletics session — have the children estimate how far they can jump and find out how accurate they were. Record and display findings. The children may like to compare their results with those of another age group.

Over-size:
★ Compare the sizes of Jack and the giant in *Jack and the beanstalk*. How many Jacks make

up the giant in height? The illustrations from a published version can be used, or the children can decide this for themselves.

★ If a small child in the grade represents Jack, find out how tall the giant will be in terms of the school building or trees in the schoolyard. Mark out Jack's and the giant's outlines on the asphalt. How many times longer is the giant's outline (perimeter)?

★ Cover the areas within the outlines with some unit (e.g. books, MAB flats, paper squares). Ask the children how many times bigger is the giant's area.

★ If the BFG is seven Sophies tall, how many Sophie footprints will fit into the BFG's footprint? Have the children draw the giant's hand and footprints, and show Sophie's prints on them.

Preps at Wonga Park made the BFG's hand and footprints.

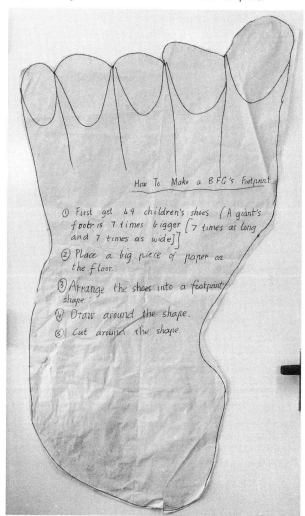

The BFG's handprint is as big as 49 children's handprints

★ Jim buys the giant a number of items. Invite the children to investigate the size of these (dimensions and mass). What would they cost in terms of current money values? What materials would be needed? How could they be made?

★ Two books about food (*The biggest cake in the world* and *The giant jam sandwich*) could lead into investigations about quantities of ingredients needed and into problems to be overcome in preparing and cooking such large items. See also the suggested activities for *Something absolutely enormous* (page 41).

★ The children can be invited to look critically at text and illustrations in terms of scale and consistency. For instance, if the BFG is seven Sophies tall, is it reasonable for Sophie to sit in his ear?

Under-size:
Similar activities to those suggested in the over-size section could be initiated. The children will probably have more difficulty with the concepts in this area than in over-size. A theme on mini-beasts is another way to explore this topic. Toys and model farm and zoo animals can also be used.

Two books which give an extended treatment of the concept are *The borrowers* and *Gulliver's travels. Arabella, the smallest girl in the world* and *The shrinking of Treehorn* suggest some problems of being tiny.

★ Allow the children to find items which could be used by the Lilliputians, the borrowers or Arabella (e.g. gum-nuts for beds or match-boxes for furniture). Older children may like to construct items for little people, or investigate questions such as how much material would be needed to make a dress for a Lilliputian or a borrower. (Would the seams need to be standard or miniature size?)

THEME: CLASSIFICATION

Introduction

This topic involves many other curriculum areas, and could easily form part of a theme or topic in science or general studies, as well as having strong links with art and language.

The activities outlined for *Grandma goes shopping* (page 30) could be used as an introductory activity for this theme. *The fat cat* is another suitable text for this, as the lists in these books are manageable, and working on one of these would give children the opportunity to discuss and compare their decisions.

Children need lots of oral practice and practice in manipulating objects in order to develop fully and articulate clearly the concepts involved in classification. Children should be involved in a variety of activities which allow mathematical concepts to be explored through language.

Great use can be made of natural material (leaves, sticks, stones, seeds etc.) and of unstructured material (e.g. buttons, wool, fabric scraps, toy animals). These can often be collected by the children at home or at school.

Children's own collections (e.g. stickers, erasers, swap cards) can also be used for classification. (See Chapter 9 for an extensive list of useful material.)

Allow time for the children to sort, classify and compare objects, to justify their sortings, make representations of their sortings and to label their objects. The children can also be encouraged to describe all the attributes of one object and compare it with others. The children have the opportunity to work individually and in groups to allow for various comparisons.

The purpose for sorting may determine the categories chosen. Sorting to share will be different from sorting by attributes; sorting children by family will have different uses from sorting them by age. The ability to classify is important in the organisation of information and hence in planning or predicting.

Booklist

The baby's catalogue, Janet and Allan Ahlberg. (**L,M**): A textless picture book following five families through the day, it has endless opportunities for discussion and classification.

Anno's counting book, Mitsumasa Anno. (**L,M,U**): There are many different items to count and classify in this no-text counting book. See page 52.

Grandma goes shopping, Ronda and David Armitage. (**M**): Grandma buys an amazing assortment of articles. See page 30.

Animalia, Graeme Base. (**M,U**): An alphabet book with sumptuous art work. A frieze is also available.

Spelling list, Michael Dugan. (**M,U**): What happens when the recipe is misread? The witch solves the problem by re-organising information.

The fat cat, Jack Kent. (**L,M**): A cumulative tale in which the cat becomes fat by swallowing an amazing assortment of items.

Clive eats alligators, Alison Lester. (**L,M,U**): A narrative about a group of children who are all different in their own special way.

'A lost button', Arnold Lobel (in *Frog and Toad are friends*). (**L**): None of the buttons his friends find match Toad's lost button. See page 19.

Mystery meals, E. Nilsson. (**L,M,U**): Children need to search the pictures in order to find the answers to the puzzles.

One dragon's dream, Peter Pavey. (**M,U**): A counting book with rich illustrations of a fantastic dream.

Mr Squire's collections, Keith Pigdon and Marilyn Woolley. (**L,M**): A book which shows in a simple manner how collections can be organised.

Sea shells, Beverley Randell. (**L,M**): A close look at the attributes of individual sea shells.

Rhymes and poems:

'Trees' (in *How big is big?*), Avelyn Davidson. (**L**): A rhyme showing different ways of describing trees.

'Dogs' (in *How big is big?*), Avelyn Davidson. (**L**): A rhyme about dogs of different shapes.

'Thick' (in *How big is big?*), Avelyn Davidson. (**L**): Different ways of using the word thick.

'Thin' (in *How big is big?*), Avelyn Davidson. (**L**): Some long and thin objects.

'Buttons' (in *Soldiers*), Avelyn Davidson. (**L**): A rhyme to stimulate discussion on ways of sorting. See page 19.

Activities

As suggested in the introduction to this theme, it is important for the children to manipulate and sort concrete material before attempting the activities based on particular books or groups of books. If new material is being used, the children will need time to explore this before attempting a set task. Allow the children to work in pairs (or small groups if they are competent at group work). Give each pair similar items to sort, list and label. Comparing with other children demonstrates that the activity can be done in different ways. Different objects may then be given to each pair or group to sort and classify. The children will probably find different ways to represent and label their sorting. These should be discussed as a class. Allow the children also to choose their own materials to classify and sort.

★ The rhyme 'Buttons' may be a stimulus for the children to sort their material in a different way. The children may choose a button or another object and list its attributes. The button story from *Frog and Toad are friends* would also be suitable for this activity. (See page 19.)

★ The rhymes 'Trees', 'Dogs', 'Thick' and 'Thin' are useful for extending and refining vocabulary. The children could build on these to find other words which describe shape and other items to fit the categories thick and thin. The rhymes could be re-written, focusing on, for example, flowers, cats, long and short.

★ After reading *Grandma goes shopping* or *The fat cat*, the children can work in small groups to classify the list of objects. This would make a good introduction to classifying the objects on a page of *Animalia, One dragon's dream, Anno's counting book*, or *Mystery meals*. Multiple copies would be an advantage, but if these are not available, each group could work on a different book. The teacher could introduce constraints such as classifying into two or three sets only. The children should be allowed to choose their own categories and should be expected to justify their decisions. They may record the number of items in each category and graph or otherwise display their results.

★ The richness and details of the illustrations in these books and especially in *Animalia* and *One dragon's dream* is such that the children will want to return to and pore over them. They will continue to find items and try to identify their names over a long period. The children may like to work on the page of their own initial from *Animalia*, and classify the items they find on this page. An associated activity would be for each of the children to create a page for their own initial. This could make a class book.

★ *Clive eats alligators* and *The baby's catalogue* both follow a group of children through a day. After reading and discussing *Clive eats alligators* the children may like to list their breakfasts, favourite foods etc., and create a class profile of their preferences. This could also be done as a small group activity, in which case the task would be to make sure that each member had a different preference in each category. The children could also, of course, make their own categories.

★ Similar activities could be based on *The baby's catalogue*. This book is also a way to use junk mail catalogues in a positive manner. The children could cut and paste items from the catalogues to make their own categories.

★ Both *Clive eats alligators* and *The baby's catalogue* are excellent for using as innovative text, or creating books for younger children.

THEME: SPATIAL RELATIONS — SHAPE

Introduction

Books offer diverse styles and presentation which give children the opportunity to explore the concept of shape in a variety of ways. The problem-solving approach we are suggesting gives new life and interest to this important topic. Many of the activities involve children in building and in manipulating material to create solutions. Others use art work. Children enjoy these kinds of activities immensely.

Booklist

Anno's magical ABC, Mitsumasa Anno and Masaichiro Anno. (**U**): An anamorphic (distorted) alphabet complete with magic mirror and instructions on creating your own anamorphic drawings.

Alex's bed, Mary Dickinson. (**M,U**): Mum and Alex solve the problem of lack of space in his bedroom — or do they?.

Changes, changes, Pat Hutchins. (**L,M**): The wooden man and woman adapt a set of blocks to suit their needs, through a series of changes from a house to a fire engine, to a boat, to a truck to a train and finally back to a house again. See page 24.

The quilt, Ann Jonas. (**L,M,U**): At night the patchwork quilt becomes a landscape.

Round trip, Ann Jonas. (**U**): The black and white drawings illustrate a journey. When you reach the end of the book, turn it upside-down and read back to the start again. See page 47.

Shapes, Jan Pienkowski. (**L**): A concept book with clear and simple illustrations, which includes less familiar shapes such as spiral and crescent.

My tower, Beverley Randell. (**L**): A simple beginner's book showing a small child building a tower.

My cat likes to hide in boxes, Eve Sutton. (**L,M,U**): Cats from other lands sing, dance or fly aeroplanes — but my cat hides in an enormous variety of boxes.

Danny's dilemma, John Tarlton. (**L,M**): Danny uses the junk in the yard to build a variety of vehicles. Compare this book with *Changes, changes*.

Animal shapes, Brian Wildsmith. (**L,M**): This is a no-text book using geometric shapes to create images of animals.

Brian Wildsmith 1 2 3, Brian Wildsmith. (**L,M**): Shape and colour combine in this dazzling counting book.

But where is the green parrot?, T. and W. Zacharias. (**L**): Each page shows a different scene incorporating a variety of shapes, and a green parrot.

Rhymes and poems:

'The song of the shapes' (in *Jack the treacle eater*), Charles Causley. (**M,U**): This delightful poem for older children describes the Misses Square, Triangle, Rectangle and Circle's visions of their future husbands and the lives they will lead. See page 63.

'Trees' (in *How big is big?*) (**L**).
'Dogs' (in *How big is big?*) (**L**).
'Thick' (in *How big is big?*) (**L**).
'Thin' (in *How big is big?*) (**L**).
'Round' (in *How big is big?*) (**L**).
'House shapes' (in *How big is big?*) (**L**).
'Shapes' (in *Ten little goblins*) (**L**).
'Tangrams' (in *Wheels*) (**L,M**).

These rhymes by Avelyn Davidson use the vocabulary of shape.

Activities

A large variety of materials such as blocks, Lego, Polydron, wood offcuts, cardboard, boxes, tangrams, attribute blocks and grid paper will be useful.

Two-dimensional shapes:

★ Use the rhymes and the Pienkowski *Shapes* as a stimulus for identifying shapes in the environment, and to extend children's vocabulary. Older children can use Charles Causley's 'The song of the shapes' as a model for their own writing about shapes in the environment.

★ Make pictures and designs using geometric shapes. The Wildsmith and Pienkowski books can be used as models.

★ Put out a number of paper squares or other shapes. The children work in pairs; the leader makes a design or picture, and the partner must 'do the same'. Ask the children to report back: are their pictures the same, and in what way? Note their comments as well as their products, as this may give some insight into their concept of 'sameness'.

★ Read 'Tangrams' and make your own tangrams. They need not be the traditional seven-piece version; a two- or three-piece version is suitable for young children.

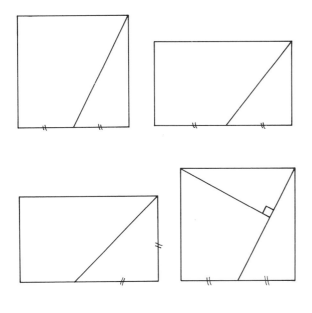

Some ways of making a 2- and a 3-piece tangram.

★ After reading *The quilt* design patchwork quilts. Try other shapes besides squares. Which will tesselate? Introduce constraints such as 'no two patches the same in any row or column'. Attribute blocks can also be used, as can different kinds of grid paper.

★ Older children may like to attempt 'upside down' drawings after studying *Round trip*. *Anno's magical ABC* gives instructions on creating your own anamorphic drawings, by transferring your drawing from a rectangular to a curved grid.

Three-dimensional shapes:
★ Reading a simple book such as *My tower* can be followed by building the tallest tower or the longest road.

★ After exploring and discussing the pictures and story in *Changes, changes,* children may use a set of blocks to make their own constructions, change them and tell a story.

★ A similar activity using junk material could follow *Danny's dilemma*. This book could also form the basis for an innovative text.

★ Discuss the boxes in the illustrations for *My cat likes to hide in boxes*. What are they for? What shapes are they? The children can collect, sort and classify boxes. Which boxes hold most? How can they measure the capacity of a box?

★ Provide cardboard for the children to make boxes. Allow them to work out for themselves how to make them. What will the children keep, or what will they hide, in their boxes?

THEME: SPATIAL RELATIONS — LOCATION

Introduction

As with shape, this section allows children to investigate the topic in formal and informal ways. Even older children who have begun to develop more formal mapping skills will enjoy, and be challenged by, creating models or relief maps of simple books such as *Go ducks go!* or *Rosie's walk*. These books allow children to use the formal language associated with location and to apply such language in a mathematical problem-solving activity.

Booklist

A lion in the night, Pamela Allen. (**L,M**): A lion 'steals' the baby; a splendid chase is followed by an equally splendid breakfast feast. See page 21.

Bears in the night, B. Berenstein. (**L**): The bears make a night journey out and back again.

The shopping basket, John Burningham. (**M,U**): On the way back from the shop, Stephen 'loses' some of his shopping. See page 31.

Go ducks go!, Maurice Burns. (**L**): Toy ducks go on a journey down a stream.

Hairy Maclary's bone, Lynley Dodd. (**L**): Hairy Maclary's friends want a share of his bone — but he manages to lose them all on the way home.

The smallest turtle, Lynley Dodd. (**L**): The story of the smallest turtle's journey to the sea.

Possum magic, Mem Fox. (**M,U**): The invisible possum travels all around Australia before the magic spell is lifted.

Rosie's walk, Pat Hutchins. (**L**): Rosie the hen walks around the farmyard oblivious to the fox behind her.

Here a chick, there a chick, Bruce McMillan. (**L**): Delightful photographs of a chick on a journey.

A pet for Mrs Arbuckle, Gwenda Smyth. (**M,U**): Mrs Arbuckle travels the world looking for a pet. See page 40.

Niki's walk, Jane Tanner. (**M,U**): Beautiful illustrations in this no-text book of a walk around Melbourne.

The crawly crawly caterpillar, H.E. Todd. (**L,M**): The life history of a caterpillar includes a journey up, along, over and under.

Activities

Have available Playdoh, Plasticine, sand or other materials for relief maps and models.

★ Use a variety of books to explore concepts of location and direction with younger children. *Here a chick, there a chick* and *Rosie's walk* will prove popular. Discuss and list vocabulary such as 'in', 'out', 'up', 'down', 'between', 'around', 'under' and 'over'. Ask the children to contribute additional words.

★ Move around (preferably outside) using the vocabulary discussed and asking children to go 'under' the slide, 'down' the pole, 'over' the A-frame, etc. Look for children who have difficulty using the vocabulary correctly.

★ Go on a trail or journey through the playground, as in *Rosie's walk*, *Go ducks go!*, and *A lion in the night*. This could also be done at home or on the way to school.

Clinton shows the street layout clearly on his map of his trip from home to the milk bar.

★ Following a reading of *A lion in the night* or *The shopping basket*, make class, group or individual maps of the journey. *Go ducks go!* would make a delightful relief map. The accuracy and detail they show will depend on the age, experience and development of the children.

★ Discuss *Niki's walk*. Where did she go? How long did it take? Write a text to match the illustrations. Use a street directory to find a possible route for Niki, or to make up a walk of your own.

These prep. and grade 1 children show great variation in the presentation and in the detail of their maps.

★ For more formal mapping, read *Possum magic* and *A pet for Mrs Arbuckle*. Locate the places on a map or atlas, and mark them on an outline map. Find the distances travelled. Has Mrs Arbuckle been efficient? Make her journey more efficient. What would her journey cost? How long would it take her? (Airline schedules would be useful.) How long will Mr A. have to cope on his own?

7:

COUNTING BOOKS

Could you have tea for two without the two
— or three blind mice without the three?
Would there be four corners of the earth if
there weren't a four? And how could you
sail the seven seas without a seven?

(The phantom tollbooth, p.149)

INTRODUCTION

A wide variety of counting books is readily available, ranging from the very simple (e.g. Dick Bruna's *I can count*) to the complicated and complex (e.g. *Anno's counting book*). Counting books illustrate many aspects of number, order and classification. Some books have a theme (e.g. Rodney Peppé's *Circus numbers*), or a narrative (e.g. Shirley Hughes' *When we went to the park*). Some show sets of totally unrelated objects through the book (e.g. Richard Hefter's *One white crocodile smile*), others count the same item on each page (e.g. Beverley Randell's *Ten small koalas*). On each page, the items may be identical, or nearly so (e.g. Pat Hutchin's *1 hunter*), or of the same kind but quite diverse (e.g. Shirley Hughes' *When we went to the park*). Ordinal number is shown in counting rhymes such as *Five little chicks* (illustrated by Wendy Hodder). Using this variety of approaches to number assists children to develop broad concepts of number and to abstract the essential features of number.

Below we list some ways of using counting books, and an annotated list of suitable titles. There are many other counting books which may be used in place of some of these titles.

ILLUSTRATING CONCEPTS

Anno's counting book, Mitsumasa Anno. (**L,M,U**): Anno uses the numbers from zero to twelve to count a number of objects, to denote the sequence of months in a year and to show the seasons, to measure time on a clock and show the passage of years, and as labels (as in a house number). See page 52.
Numbers, Jan Pienkowski. (**L**): This simple book shows the numbers from 1 to 10 as both ordered arrays and random collections.
One duck, another duck, Charlotte Pomerantz. (**L**): Counting is shown as adding another one to the collection each time. Large numbers are introduced at the end as Danny the owl thinks he will count the stars.

WRITING

Thematic and narrative counting books lend themselves to children creating their own books using these as models. Younger children can practise counting as they write and illustrate their books. All children will be challenged by meeting the particular criteria in a model whether theme, narrative, or form. Older children may like to make counting books for younger children, or help younger children by scribing for them. A number of counting books can be presented together, so that children can choose which features to include in their texts, or the teacher may wish to use a particular text for a particular purpose. There are many counting books besides those listed here which would make good models for children's writing.

Ten go hopping, Viv Allbright. (**L**): One little boy goes hopping — followed in turn by a grasshopper, a mouse, a frog, a rabbit, a cat, a dog, a monkey, a kangaroo and an elephant. This is a good book to act out in the playground. Discuss the increase in size of the animals.

Ten, nine, eight, Molly Bang. (**L**): A delightful countdown to bed-time starting with '10 small toes all washed and warm' and ending with '1 big girl all ready for bed'. Many children have difficulty counting backwards (except when launching a rocket), so try adapting this to a countdown for Christmas, or for a birthday party. Alternatively, ask the children to use their own bedrooms, houses or gardens for a counting book. They will need to choose items of which there are exactly 10, 9, 8 and so on down to 1. See page 13.

When we went to the park, Shirley Hughes. (**L**): The story of a little girl's walk to the park with her grandfather. Notice in the illustrations how different each of her 'six runners running' are. Look also for things of which there are so many they can't all be counted. See page 16.

1 hunter, Pat Hutchins. (**L**): The hunter doesn't notice all the animals until the end, when he flees. In contrast to *When we went to the park*, the animals are highly stylised and alike. See page 18.

Over in the meadow, Olive A. Wadsworth (Puffin, 1985)

Over in the meadow, Ezra Jack Keats (Hamish Hamilton, 1973). (**L,M**): There are many versions of this traditional rhyme, which lends itself to adaptation to an Australian setting — 'Over in the paddock'.

Toot toot, Brian Wildsmith. (**L**): No numbers appear in this animal counting book; the numbers are shown in the pictures and in the animal noises. Besides making a new text, the children could construct the train and its wagons shown on the final page.

COUNTING

By their nature, all counting books can be used for counting. Those listed here have some special features:

Anno's counting book, Mitsumasa Anno. (**L**): There are many items to count on each page. See page 52.

When we went to the park, Shirley Hughes. (**L**): Look for things so numerous you 'couldn't count them all'. Older children can devise ways of counting things such as grains of rice, or leaves. See page 16.

Numbers of things, Helen Oxenbury. (**L**): Most counting books go to ten or twelve; this one goes to 50 ladybirds, and ends with 'how many stars?'

One dragon's dream, Peter Pavey. (**M**): Older children will enjoy poring over the rich illustrations, and counting the objects in each set.

Numbers, John Reiss. (**L**): Reiss claims 1000 raindrops on his final page.

CLASSIFICATION

The objects in a counting book can be classified according to whatever criteria the children (or teacher) decide. The number in each group can then be found and compared. Children can use various methods of describing, representing and recording their groupings. They should be prepared to justify their system of classification. Many counting books besides those listed here can be used for classification. See also the section on **Classification**, Chapter 6.

Anno's counting book, Mitsumasa Anno. (**L,M,U**): The items on each page can be classified. See page 52.

Anno's counting house, Mitsumasa Anno. (**L,M**): Ten children move house in this no-text book, taking things with them as they go. The book affords great scope for sorting and classifying. Look at the children's clothes and sex, as well as at their possessions.

One white crocodile smile, Richard Hefter. (**L,M**): A diverse collection of objects in this simple counting book.

COMPUTATION

Many counting books and rhymes, besides those listed here, can be used for addition and subtraction.

One two three going to sea, Alain. (**L**): A cumulative tale which shows addition and subtraction to 10.

Anno's counting house, Mitsumasa Anno. (**L**): As the ten children move, one by one, to the new house, addition and subtraction equations can be made. If there are three in one house, there must be seven in the other. $3 + 7 = 10$, $10 - 3 = 7$.

1, 2, 3 to the zoo, Eric Carle. (**L**): Find how many animals are on the train to the zoo.

Teddybears 1–10, Susanna Gretz. (**L**): How many teddybears are there altogether in this counting book? Use drawing, concrete material or tallying to find out.

Ten apples up on top, Theo LeSieg. (**M**): This zany counting book lends itself to comparison of numbers, addition and subtraction as the difference between numbers. Children can make up equations or other number sentences about a particular page; other children can identify the corresponding page. See page 35.

Ten little swimming crabs, Beverley Randell. (**L**): 'Ten little swimming crabs found a fishing line. One caught it with his claw and then there were nine.' When only one crab is left, it meets nine more baby crabs. Use this book to illustrate subtraction. The children can represent the number change at each stage in their own ways — pictorially, by tallying, on a number line, with concrete material or in equation form.

Over in the meadow, Olive Wadsworth (Penguin, 1986)

Over in the meadow, Ezra Jack Keats (Hamish Hamilton, 1973). (**L**): How many animals were there in the meadow? Remember to count in the mother animals.

PATTERN AND ORDER

Triangular numbers can be studied through counting books. The books listed lend themselves to pictorial representation of the triangular pattern. Older children may like to find a quick way of summing the numbers from 1 to 10 or 12, and then extend this to summing, say, 1 to 100.

At the zoo, Beverley Randell. (**L**): An Australian counting book of a trip to the zoo. Represent the animals seen in a triangular array.

The twelve days of Christmas, Brian Wildsmith. (**L,M**): Arrange the gifts in the shape of a Christmas tree to see the pattern. See page 43.

8:

ANNOTATED

BOOKLIST

OF OTHER

MATHEMATICAL

TOPICS

If a small car carrying three people at thirty miles an hour for ten minutes along a road five miles long at 11.35 in the morning starts at the same time as three people who have been travelling in a little automobile at twenty miles an hour for fifteen minutes on another road exactly twice as long as one half the distance of the other, while a dog, a bug, and a boy travel an equal distance in the same time or the same distance in an equal time along a third road in mid-October, then which one arrives first and which is the best way to go?

(The phantom tollbooth, p. 147)

INTRODUCTION

In this chapter we list books under some further topic headings, with some brief suggestions on possible activities.

OPERATIONS AND COMPUTATION

One two three going to sea, Alain. (**L**): A cumulative tale of too many fishermen in too small a boat. Practise addition first and then subtraction.

Two hundred rabbits, L. Anderson. (**M,U**): Whilst 199 rabbits can't be arranged in an array of rows and columns, 200 can. Start an investigation into composite and prime numbers, and factors.

Anno's counting house, Mitsumasa Anno. (**L**): Investigate numbers that add to 10 with this story of ten children moving house.

Anno's mysterious multiplying jar, Mitsumasa Anno. (**U**): Multiplication, factorials and large numbers can be explored in this magical tale of a jar. See page 46.

Noah's Ark song, June Epstein. (**L**): Multiplication by two is explored in this Australian song.

Phoebe and the hot water bottles, Terry Furchgott and Linda Dawson. (**M**): Phoebe has 158 hot water bottles when she is eight. Make combinations of these to show how many she was given each year. See page 33.

A bag full of pups, Dick Gackenbach. (**L,M**): Practise recording subtraction with this story of giving away twelve pups. See page 23.

Mr Mullin has nine pups left because the farmer took one to herd his cows and the magician took two to do tricks with them.

the farmer took one pup.   the magician took two puppies.

There are nine pups left to give away.

$12 - 8 = 4$

Grade 2/3 children at Cooinda show different methods of representing the number of pups in the story *A bag full of pups*.

Preps at Camberwell made drawings to show how to share out cookies evenly, after reading *The doorbell rang*.

The doorbell rang, Pat Hutchins. (**L,M**): A story of sharing out cookies between twelve children. See page 26.

King Kaid of India, (Victorian Fifth Reader). (**U**): The inventor of chess demands a reward which involves doubling and reaches astronomical proportions. See page 48.

Bunches and bunches of bunnies, Louise Mathews. (**M**): A rhyming book showing the squares from 1×1 to 12×12.

Mr Brown's magnificent apple tree, Yvonne Winer. (**L**): Practise subtraction with this story of a mouse family and an apple tree.

MASS, WEIGHT AND VOLUME

Mr Archimedes' bath, Pamela Allen. (**L,M,U**): Investigate why the bath water goes up and down as you get in and out. See page 50.

Who sank the boat?, Pamela Allen. (**L,M,U**): Questions of balance, floating and sinking are raised in this story of animal friends going out in a boat. See page 51.

Dad's diet, Barbara Comber. (**M,U**): Dad manages, painfully, to lose weight — from 84 kilograms to 72 kilograms. He ends up weighing $6/7$ of his former self. Find out what the other members of the family weigh. Investigate the weights of the class, of other grades in the school and of younger brothers and sisters. Find relationships between these weights — e.g. can the children find someone whose weight is ½ theirs? 1½ theirs? See page 37.

The light princess, George MacDonald. (**M,U**): A fairytale about the princess who is without gravity. She laughs a lot, and has trouble coming down to earth. Good for discussion of the difference between mass and weight, and of the problems of weightlessness.

Melisande, E. Nesbit. (**M,U**): Melisande's hair grows long and thick — and even longer and thicker each time it is cut off. The prince tries to solve the problem by cutting the princess off the hair instead of the hair off the princess; but then Melisande grows. Equality solves the problem. This is another story to provoke thought and discussion.

The great big enormous turnip, Alexei Tolstoy. (**L,M**): It takes the old man, the old woman, the granddaughter, the dog, the cat and the mouse to pull up the great big enormous turnip. Compare the story with *Who sank the boat?*, where it is apparently the mouse who sinks the boat. Is it the mouse who pulls up the turnip? Are the biggest and heaviest people the best at tug-of-war? Discuss, and experiment.

TIME

All in a day, Mitsumasa Anno. (**M,U**): Anno and eight other well-known illustrators have contributed to this book. Each page shows nine children around the world and what they are doing at the same time. Using a globe, discuss why the time, seasons and climate are different for the countries depicted. The children could make their own drawings and text for their own day. Choose another country (perhaps one where a class member has friends or family) and show a typical day for a child there alongside their own day. Show the time in both countries. See page 44.

Aidan shows his night- as well as day-time activities in his picture of *All in a day*.

Anna's day, Peggy Blakely. (**L**): Activities through the day for a little girl are shown in this book. A different kind of clock shows the time on each page.

The bad-tempered ladybird, Eric Carle. (**L,M**): Besides showing the clock face through the day, the pages of this book show the movement of the sun. Go outside and investigate the height and direction of the sun through the day. Compare summer and winter.

The very hungry caterpillar, Eric Carle. (**L**): One week in the life of a caterpillar. Practise the names and sequence of the days. See page 14.

The new baby calf, Edith Chase. (**L,M**): The illustrations to this poem show the changing seasons as the little calf grows.

Our teacher Miss Pool, Joy Cowley. (**L**): Miss Pool arrives at school by a different and unusual means of transport each day of the week.

Boss for a week, Libby Handy. (**L**): There is a new rule for each day of the week — for everyone except little Caroline. Note the calendar as well as the names of the days.

What's the time Mr Wolf?, Colin Hawkins.
What's the time Mr Wolf?, Elizabeth Honey.
What's the time Mr Wolf?, Beverley Randell. (**L**): Use any of these versions before playing the game. Discuss the clock time and what happens at those times. Better still, compare the versions and ask children to make their own version, showing what they are doing at different times.

Lucy and Tom's day, Shirley Hughes. (**L**): Delightful illustrations to this description of two small children's day.

Clocks and more clocks, Pat Hutchins. (**L,M,U**): As he moves from room to room, Mr Higgins' clocks all say different times. The clockmaker solves the problem with his wonderful watch. Try a trip around the school with and without a watch. Or ask children to represent Mr Higgins' problem in another way.

Rhymes around the day, Jan Ormerod. (**L**): Identify the time and occasion for each rhyme in this collection.

Moonlight, Jan Ormerod. (**L**): A no-text look at a family at night.

Sunshine, Jan Ormerod. (**L**): The day-time companion to *Moonlight*.

A day on The Avenue, Robert Roennfeldt. (**M,U**): A day in a Geelong street. See page 39.

MONEY

Going shopping, Sarah Garland. (**L**): A realistic look at shopping with children and a dog! Use the book to introduce a unit on money.

Teddybears go shopping, Susanna Gretz. (**L**): Shopping in a supermarket. This is a good starting point for investigations in a number of areas, including money. See page 15.

Toby's millions, Morris Lurie. (**M,U**): Money doesn't necessarily bring happiness. How would you spend millions? Check that your costing is realistic.

Going shopping, Greg Mitchell. (**L,M**): See what this family buys when everything costs too much.

Alexander, who used to be rich last Sunday, Judith Viorst. (**M**): Alexander was rich last Sunday when he had two dollars... but now all he has is toy money. Ask the children how they would spend two dollars, or $100. See page 36.

9:

CLASSROOM

ORGANISATION

AND OTHER ISSUES

You must never feel badly about making mistakes... as long as you take the trouble to learn from them. For you often learn more by being wrong for the right reasons than you do by being right for the wrong reasons.

(*The phantom tollbooth*, p. 196)

In this chapter we present some of the current thinking on children's learning, the implications of that thinking for the mathematics classroom, and the most effective use of the suggested activities in this book.

HOW CHILDREN LEARN MOST EFFECTIVELY

A number of factors which appear to promote effective mathematical learning have been identified by researchers such as Ed Labinowicz (1985), Martin Hughes (1986) and Margaret Donaldson (1979). Those we include here seem most relevant to this book, though of course our choice is neither exhaustive nor definitive.

Active involvement

Children should be actively involved in their learning. This means not only physical manipulation of materials and physical movement by the child, but also includes planning, implementing and reflecting. Children who are actively involved show the 'steadfast intent' described by Terry Johnson. Difficult to describe but easy to recognise, this 'steadfast intent' is likely to occur when 'the individual feels autonomous, competent and motivated' (Johnson: 1989, pp. 26-7).

Social interaction

Children, like adults, are social beings and the sharing of ideas, achievements and failures is important. Such sharing will affect both attitude and cognitive development.

Talking

Talking about their work helps children to clarify their thinking, to learn from each other, to justify their findings and to share the joy of discovery and achievement with others. Talking is important in all aspects of active involvement — planning, implementing, reflecting.

Experimentation

Children should be prepared and encouraged to generate and try out new ideas. They should have the confidence to take risks, and from their experimentation be able to develop new understandings or further investigations.

Learning from errors

Errors are often a point of learning. A child who is aware of error is challenged by a dissonance between her or his theory and experience, and in reconciling this dissonance will develop a more effective theory. This process will assist in the future when the child is applying the theory in another situation. (Not all errors can be regarded in this way of course. However, even many of the errors which appear at first sight to be merely careless or random mistakes may in fact reveal a limited or confused understanding.)

Feedback

Feedback helps children to investigate and correct errors or to confirm their thinking. The feedback is needed whilst the child is still actively involved in the task, and should be of a kind which has meaning for the child and which assists in developing mathematical thinking. Discussion with others and comparison of methods and results is one way of gaining feedback.

Practice

Children need opportunities to practise new skills in a meaningful context. Repeating the same activity with different materials, and doing similar activities in a different context helps to consolidate skills as well as developing fuller understanding.

Reflecting

Children need time to reflect. It is through reflection that they determine further action and develop generalisations and theories. Discussing work is an important component of reflection. Writing about their work can also help children to clarify their thinking.

IMPLICATIONS FOR THE CLASSROOM

In order to make learning most effective, teachers will need then to set up conditions which make best use of the principles we have outlined. This will mean providing suitable physical and social environments and appropriate challenges for the children.

Physical environment

Organisation of space:

The arrangement of the room should be adaptable to a variety of activities and groupings. Children need to be able to move around, for instance to collect materials, without disturbing others, and they need to be able to discuss with others without shouting or otherwise disrupting the rest of the class. Where space is limited some ingenuity may be needed. Going outside may solve some problems of space, noise and mess, and many of our lessons have an outdoor component. It will also ensure the teacher's popularity! (See Goodnow: 1985, pp.106–7).

Materials:

Equipment of various kinds should be provided. These should include both structured and unstructured materials, games (bought or home-made), paper and card, paste, masking tape and scissors. It is more important to provide a variety of materials than to provide a large amount of any particular item. Doing the same or similar activities with different materials gives children the practice they need besides helping them to develop concepts more fully. Children need ready access to materials, and in general free choice of materials. However there may be occasions when teachers wish to restrict choice of material.

In general, it is wise to introduce new material to children some time before doing the particular activities as they need time to explore the possibilities of the material. The teacher who produced Plasticine specially for *The very hungry caterpillar* found that the children were enthralled with the new material, and not interested in the task she had set.

The following equipment may be useful:
- abacus (spike and loop)
- acetate sheets
- acorns
- attribute blocks
- balances
- bathroom scales
- beadframe
- beads
- blocks (all shapes and sizes)
- bolts
- bottle tops

- bottles (all shapes and sizes, preferably plastic)
- boxes (all shapes and sizes)
- buttons
- calendars
- cardboard
- clocks (all kinds)
- clothes pegs
- coins (plastic, cardboard, stamps)
- coloured paper squares
- construction kits
- containers
- corks
- counters
- Cuisenaire rods
- cups
- clay
- egg cartons
- fabrics (all sorts and patterns)
- feathers
- farm and zoo animals
- foam rubber
- funnels
- fuzzy-felt shapes
- geoboards
- graduated containers
- graph paper (all kinds)
- icy-pole sticks
- interlocking cubes
- kitchen scales
- lids (all sizes)
- MAB
- marbles
- match-boxes
- mirrors
- nails
- nuts and seeds (all sizes)
- pasta
- pattern blocks
- peg-board and pegs
- pipe cleaners
- Plasticine
- Playdoh
- playing cards
- ribbons
- rice
- rubber bands
- rulers
- sand
- sawdust
- shells
- spring balance
- stones
- straws
- string
- tangrams
- tape measures
- toothpicks
- wallpaper and wrapping paper
- water tray

Social environment

The atmosphere of the classroom is probably the single most important factor in establishing a good working environment. It is however a difficult area to discuss as the atmosphere is largely determined by the teacher's attitude. Nevertheless, the following suggestions may help both in establishing and maintaining an encouraging atmosphere.

Valuing children's contributions:
Children need to feel that their contribution and ideas are accepted both by the teacher and by each other. Only if they feel this acceptance will they be prepared to take risks and thus learn from their errors as well as their successes. Oral discussion, both child–child and child–teacher, and reporting to the whole class — to a supportive but not undiscriminating audience — are important components of these lessons. The requirement that children justify their workings and understandings helps to establish this accepting atmosphere as well as developing their thinking about mathematics and its applications. Of course, both teacher and children should develop a sensitivity to each other in order to give positive feedback and constructive comments whilst challenging children's misconceptions and misunderstandings.

Groupings:
Because different children have different preferred learning styles, and because the ability both to work independently and in a group is important in later life as well as at school, we recommend a variety of groupings. This does not mean that children should change frequently from one kind of grouping to another, but rather that over the term or the

year children should experience different groupings. Many of our activities suggest small groups or pairs working together, and in most cases we intend these to be co-operative groups — that is, groups in which there is a shared goal and a single product rather than the kind in which each child has an individual goal and makes an individual product.

Discussion from our working
session on co-operative learning

Good Things

working together
co-operating
fun
not working with best friends
girls and boys mixed
did sums together
it was Maths
lots of work

Need to improve

not to work by ourselves we are a
group

listen to each other
talk to each other.
we do need to co-operate.

Grade 3 at Wonga Park summarised their discussion on co-operative learning.

In working co-operatively, children develop both intellectual and social skills. Joan Dalton (1985) gives valuable guidance on setting up co-operative learning. She emphasises the need to hasten slowly, and to teach consciously the behavioural skills which children will need in order to operate effectively. If children are not used to working in co-operative groups, an extended period working in pairs may be appropriate.

At times during these activities children will be working in loose groupings rather than co-operative groups. For instance, if the children are recording the number of pieces of fruit the very hungry caterpillar ate, the teacher may decide to set up different materials at different stations around the room. In this case, children working at any one station will have individual goals and products. The sharing they do as they work will still be important, but they will not form a co-operative group.

Groups can be chosen in a number of ways. In general, mixed ability and randomly selected groups have the greatest potential for learning. However, with particular classes or for particular purposes ability, interest or friendship groups may be appropriate. The needs of particular children and of the class as a whole may also influence choice of groupings. Dalton (1985, pp.19–20) discusses the needs of the quiet child, the child with poor verbal expression, the dominant child, the disruptive child and the loner.

Time to read:
Children need time to become familiar and to feel comfortable with books before becoming involved in activities. As well as the teacher reading to the children, books used should be available for the children to read and browse through over a period of time. Some books need close examination in order to identify the mathematics and to realise the full potential of the activities.

Time to talk:
Before children embark on an activity they may need time to clarify the task, and to plan their method, their preference for presentation and their choice of material. In a very open-ended activity such as creating a problem at the end of Margaret Wild's *Something absolutely enormous* they will need time to consider the possibilities available. Talking plays an important part in all this. As children worked on the activities we found that they discussed in groups and exchanged ideas before putting pen to paper and working individually. Time is needed to plan a method of presentation, as in Terry Furchgott and Linda Dawson's *Phoebe and the hot water bottles* or *The Hilton hen house* by Jo Hinchcliffe. Children need time to discuss their choice of material as in Eric Carle's *The very hungry caterpillar* or the poem *Band-Aids* by Shel Silverstein. They also need to talk as they are doing the activity —

to check how they are progressing and gain feedback from each other. At the end of the investigation, children need to talk about what they have achieved, and to listen to each other's

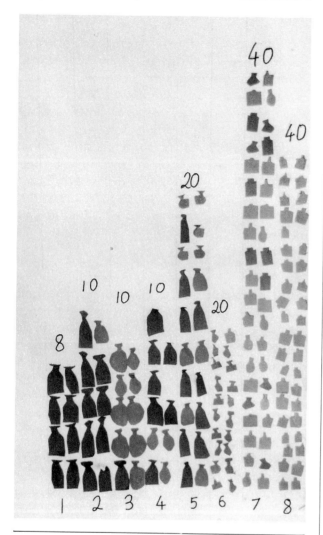

Children at Waverley Park and Wonga Park chose different methods of presentation for *Phoebe and the hot water bottles*.

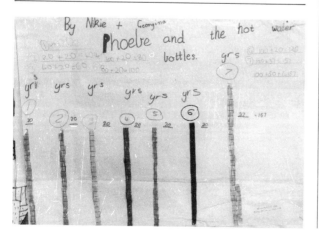

comments. Through this sharing activity children receive positive reinforcement and also have the opportunity to learn from one another. A variety of audiences is needed — the whole class, small groups, older children, the teacher or other adults.

Time to do:
Children need time to work through and complete their investigations. We found that children often needed longer than we had allowed to do the activity thoroughly, and that they often wanted to repeat the activity with other materials or to correct errors or misunderstandings. After children have shared the results of their investigations, they are often motivated to repeat an activity, or to develop it further.

Time to share:
When children have completed an activity, whether it has been done individually, in pairs or in groups, it is important for time to be set aside to share and compare results. This allows children to see what others have done, and develop and refine their ideas about the activity. It is a valuable time in which to develop oral language and critical evaluation skills. The teacher can observe the use of mathematical language and reporting skills. By participating in such reporting children are actually learning from one another, a reminder that the teacher is not the only one who teaches in the classroom.

Appropriate challenges

The story context for the problems in this book gives the activities meaning for the child. Many are open-ended, and all allow for a variety of methods of attack, choice of materials or choice of presentation. In some, children devise the problems or investigation for themselves, enabling them to respond at their own level of competence. If an initial activity proves simple for a child or group, it can be modified or extended, either by the teacher or by the children themselves. For instance, the seven-year-olds who decided that Grandma brought twelve cookies in *The doorbell rang*, and therefore finished the task quickly, were pleased to be challenged further when asked to find out what would happen if Grandma had brought lots of cookies. Some chose then to try 80.

The suggestions made in this chapter are focused on the activities and the approach this book takes. However, they reflect current thinking about mathematics education, and have wider applications. Teachers who decide to use the school environment or their science or social education curriculum as a focus for teaching mathematics will need to consider the same issues.

10:

ASSESSMENT

AND EVALUATION

'Did you know that if a beaver two feet
long with a tail a foot and a half long can
build a dam twelve feet high and six feet
wide in two days, all you would need to
build the Kariba Dam is a beaver sixty-
eight feet long with a fifty-one foot tail?'
'Where would you find a beaver as big as
that?'
'I'm sure I don't know, but if you did,
you'd certainly know what to do with him.'

(The phantom tollbooth, p.148)

INTRODUCTION

In this chapter we outline a variety of assessment and evaluation techniques suitable for use with the mathematical activities presented in Chapter 4. The indicators at the end of each activity used in conjunction with this chapter will assist in child assessment and lesson evaluation.

Such information assists the teacher in contributing to an overall profile of each child, and at the same time evaluating the unit undertaken. Such data will contribute to, and can be integrated with, the school assessment and evaluation policy. The Mathematics Curriculum and Teaching Program kit, *Assessment alternatives in mathematics* (1988) may be of use.

ASSESSMENT — WHY?

The prime purpose of assessment is to improve children's learning. In order to achieve this goal teachers need to use strategies which will elicit as much information as possible about each child. This information may be used by the child, teacher and parent to assist future learning, and in the teacher's case to plan future learning experiences. If assessment techniques are to be successful it is important that the gathering of information is as accurate as possible and that the results are used effectively in determining future lesson planning. Assessment allows teachers to take note of both strengths and weaknesses. Teachers can then build on the strengths of the child and also allow for practice and reinforcement in areas of weakness.

WHAT TO ASSESS

Problem solving

Some examples of questions which teachers may ask in order to assess children's use of problem solving strategies may be:
- Do the children write and organise lists?
- Do the children draw diagrams and label them?
- Do the children model using concrete materials?
- Do the children talk through the problem with another person?

- Can the children make up their own problems?
- Do the children check with alternative materials?
- Do the children seek alternative ways to solve problems?
- Do the children use a variety of recordings?
- Do the children work backwards rather than always start from the beginning?
- Are the children able to explain strategies to each other?

These observations not only give teachers insight into mathematical development but also allow other children to learn by methods and strategies being explained.

Skills

Many of the lessons give children the opportunity to practise mathematical skills in a meaningful context. This is important because it offers the teacher a number of ways to present similar mathematical concepts. Skills which are being assessed will depend on the activity and on the children's stage of development. Lists of skills may be found in curriculum documents and text books. Questions indicating some of the kinds of skills which may be assessed are:

- Do the children count accurately?
- Do the children know their tables?
- Can the children use their rulers to measure accurately?
- Can the children extract information from a graph?

Children who are coping well can be given extension activities and additional challenges, and those needing reinforcement activities may repeat the activity with different materials.

Understanding of concepts

By involving children in the type of activities suggested teachers can observe whether the children understand concepts and can apply skills in different situations. Comparison of the skills noted above with the related concepts which follow will assist teachers in clarifying their thinking when observing children. Some questions which teachers might ask to ascertain whether the children understand concepts rather than apply skills in a rote fashion are:

- Do the children count in groups rather than count by ones?
- Do the children recognise patterns, for example the relationship between the two- and four-times table?
- Do the children know when to use multiplication?
- Do the children understand the difference between length and area?
- Can the children draw an appropriate kind of graph to present information?

Use of language

With all the activities presented teachers need to be constantly aware of children's language. When children talk to the teacher, talk to each other or even when they are talking to themselves they are revealing what they know or what they may need to know about mathematics. Using similar activities and different concrete materials allows children to practise mathematical language in a variety of situations. Such practice assists with the internalisation of language and mathematical concepts. Teachers can record the development of formal mathematical language when they are gathering information.

Individual and group behaviour

While children are involved in activities teachers can assess both individual and group behaviour. The ability to work individually and in a group is important both in itself and because it assists in developing mathematical skills and understandings. Teachers can identify children who may be having difficulties by observing their behaviour and their contribution to the activity. And through observation, listening and talking the teacher can assess the listening and questioning skills of the children. Social skills such as being courteous, taking turns and sharing material can also be assessed (Dalton: 1985, p.182).

Attitudes

Teachers need to observe children's interaction with others in order to make a judgement about attitude. Individual observations, such as confident actions, enjoyment of task, helping others etc. can reveal to the teacher problems or successes a child may be having. For example, reluctance to begin a task may indicate that the child lacks confidence in mathematics. The teacher may need to observe over a period of time and talk to the child frequently in order to identify the problem. During mathematical activities the teacher can observe the attitudes of children and assess whether they are prepared to take risks, experiment with other material, or are taking interest in the task.

GATHERING AND RECORDING INFORMATION

We believe that the gathering and recording of information needs to be as thorough as possible, but at the same time methods must be effective and efficient to assist the busy teacher. One way to try solving this problem is to experiment with a variety of methods and to choose those which appear most useful but manageable in terms of time and effort.

Class lists

Class lists can be used as checklists (for example, can all children count to 20?) or as a means of collecting short anecdotal records (for example Sally counted correctly to 100 today).

Samples of work

Work samples are a valuable source of information and they have various uses. Samples are good resources for report writing as well as for showing parents and colleagues. Over a period of time the development and change in children's work can be observed. This is helpful to teachers and parents as well as allowing children to observe and enjoy evidence of their progress. A scrap-book can be used to keep samples showing evidence of new skills or concepts. Sometimes children are reluctant to part with work and the teacher may have to photocopy the sample.

Samples of group work can be collected in the same way to allow teachers to focus on how children work through a task and achieve results.

Observation

Observation may be carried out in a number of ways and is a valuable source of information.

General observation of the classroom and how children interact can assist the teacher in planning. For example, if group work is planned children need to be able to talk and work with one another. Observation can focus on one child or on a group of children. The teacher may choose to observe a group from the beginning to the end of an activity to gain insight into how the group interacts, approaches a task and achieves results. On the other hand the teacher may choose to focus on a particular skill, telling the children, for example, 'Today we are going to pay attention to our counting'. If this is shared with the children it also becomes a form of self-evaluation and assists them in developing that skill.

Talking and listening to children

Talking and listening to children is closely linked to observation. Sometimes the teacher may simply observe and listen while at other times talk will be included. The importance of talk cannot be over-emphasised as the teacher can assess children's mathematical thinking and their application of mathematical skills through their use of language. This is particularly important when children are justifying their methods and results.

When the teacher listens to a group at work the personalities of the group and the interaction of the children may be observed. Some aspects of that interaction may be group decision-making, the way in which problems are approached, and how children interact with one another.

EVALUATION OF SESSIONS

Combined with the indicators at the end of the activities the following questions may assist the teacher in evaluation of the session.

- Were all the children actively involved in the mathematical task?
- Was there evidence of enjoyment and satisfaction from the task?
- Did the children persevere with the task and seek alternative solutions?
- Did the children value the work they produced?
- Were the children willing to take risks, experiment, and reassess the problem?

- Did the children talk the problem through with others and refer to them while working?
- Were the children articulate when reporting to others?
- Was the work clear enough to be interpreted or read by others?
- Did the children use formal or informal language when discussing their mathematics?
- Did the children use formal or informal methods of recording while engaged on the task?
- Did the children achieve the mathematical objective of the lesson/unit?
- Could the children read the instructions if necessary?
- Could the children follow the instructions?
- Did the children listen to each other while reporting?
- Are the children developing skills in evaluating other children's work?
- Are the children applying known mathematical skills in the task?
- Were the instructions clear and unambiguous?
- Did the children have easy access to needed materials?
- Have the children been given sufficient time for discussion: of planning, of executing and of reflecting?

CONCLUSION

Teachers should not use all these evaluative questions at the same time, but rather we suggest they choose questions to suit the particular activity being evaluated. Teachers may choose a focus, and over time build up a case study. Such an approach gives order and meaning to teacher observation.

The types of assessment and evaluation techniques suggested here are linked to the kinds of activities we suggest in Chapter 4 and are for a classroom which values co-operation, a variety of activities, challenges and excellence; a classroom in which all aspects of the child are valued and a holistic approach towards assessment is taken; a classroom with a positive working atmosphere and a commitment to developing a love of mathematics and literature.

BIBLIOGRAPHY AND INDEX

The books listed may be used in a mathematics program. The list may also be used as an index to this book. Those in bold type are treated in detail. As in the text, the age range for each book is indicated by **L** (lower primary), **M** (middle primary), **U** (upper primary) and **PP** (post primary). Dates of publication are those of the editions which we used in compiling the list. Where possible, we have listed paperback editions.

Key to levels: L = lower primary, 5 to 7 years of age; M = middle primary, about 7 to 9; U = upper primary, 9 to 12.

Carle, E.,	***The very busy spider***. Putnam, 1989.	**L,M**	6, **32**
Carle, E.,	***The very hungry caterpillar***. Putnam, 1989.	**L**	10, **14**, 83, 85, 87
Carle, E.,	*1,2,3 to the zoo*. Putnam, 1989.	L	80
Carroll, L.,	*Alice in Wonderland* (many editions)	**U,PP**	
Carroll, L.,	*Through the looking glass* (many editions)	**U,PP**	
Chase, E.N.,	*The new baby calf*. Ashton Scholastic, 1986.	L	83
Comber, B.,	***Dad's diet***. Ashton Scholastic (Bookshelf), 1987.	**M,U**	**37**, 82
Cowley, J.,	*Our teacher, Miss Pool*. Nelson (Windmill), 1985.	L	83
Cowley, J.,	*Ten loopy caterpillars*. Griffin Press, 1985.	L	24
Cowley, J.,	*Where can we put an elephant?* Nelson (Windmill), 1986.	L	
Cowley, J.,	*The biggest cake in the world*. Nelson (Ready to read), 1985.	L	68, 71
Dahl, R.,	*The BFG*. Penguin, 1985.	**L,M**	68, 70
Dickinson, M.,	*Alex's bed*. Andre Deutsch, 1980.	**M,U**	9, 66-7, 74
Dodd, L.,	*Hairy Maclary's bone*. Gareth Stevens, 1985.	L	76
Dodd, L.,	*The smallest turtle*. Gareth Stevens, 1985.	L	76
Dugan, M.,	*Nonsense numbers*. Ashton Scholastic, 1982.	L	
Dugan, M.,	*Spelling list*. Macmillan. (Southern Cross), 1987.	**M,U**	72
Elkins, B.,	*Six foolish fishermen*. Hodder and Stoughton, 1983.	L	
Epstein, J.,	*Noah's Ark song*. Macmillan (Southern Cross), 1987.	L	81
Fox, M.,	*Arabella, the smallest girl in the world*. Ashton Scholastic, 1986.	**L,M**	68, 71
Fox, M.,	*Possum magic*. HBJ, 1990.	**M,U**	23, 76-7
Furchgott, T., and Dawson, L.,	***Phoebe and the hot water bottles***. Picture Lions, 1977.	**M**	**33**, 81, 87
Gackenbach, D.,	***A bag full of pups***. Ticknor & Fields, 1983.	**L,M**	**23**, 81
Gag, W.,	*Millions of cats*. Putnam, 1977.	**L,M**	46
Garland, S.,	*Going shopping*. Puffin, 1985.	L	83
Garner, A.,	*The owl service*. Collins, 1967.	**U,PP**	
Gelman, R.,	*The biggest sandwich ever*. Scholastic, 1980.	L	
Green, J.F.,	*There are trolls*. Ashton Scholastic, 1983.	L	
Greenwood, T.,	*V.I.P. Very important plant*. Puffin, 1975.	**L,M**	28
Gretz, S.,	*Teddybears 1-10*. Macmillan, 1976.	L	9, 16, 80
Gretz, S.,	***Teddybears go shopping***. Macmillan, 1982.	**L**	10, **15**, 83
Gretz, S.,	*Teddybears' moving day*. Macmillan, 1988.	L	16
Handy, L.,	*Boss for a week*. Scholastic, 1982.	L	83
Hawkins, C.,	*What's the time, Mr Wolf?* Collins, 1986.	L	83
Hefter, R.,	*One white crocodile smile*. Nelson, 1983.	**L,M**	78, 80
Heide, F.,	*The shrinking of Treehorn*. Dell, 1979.	**M,U**	68, 71
Himmelman, J.,	*Amanda and the magic garden*. Penguin, 1987.	L	68
Hinchcliffe, J.,	***The Hilton hen house***. Ashton Scholastic, 1987.	**L,M,U**	**38**, 87
Hindley, J.,	*The animal parade*. Collins, 1985.	L	
Holm, A.,	*I am David*. Puffin, 1969.	**U,PP**	
Honey, E.,	*What's the time, Mr Wolf?* Macmillan (Southern Cross), 1987.	L	83
Howell, L. and R.	*Winifred's new bed*. Hamish Hamilton, 1985.	L	66-7
Hughes, S.,	*All shapes and sizes*. Lothrop, 1986.	L	
Hughes, S.,	*Lucy and Tom's day*. Penguin, 1986.	L	83
Hughes, S.,	*Lucy and Tom's 1.2.3*. Grossman, 1987.	L	
Hughes, S.,	***When we went to the park***. Lothrop, 1985.	**L**	16, 78-9
Hunkin, T.,	*Mrs Gronkwonk and the post office tower*. Angus and Robertson, 1973.	**M,U**	
Hutchins, P.,	***Changes, changes***. Macmillan, 1987.	**L,M**	10, **24**, 74-5
Hutchins, P.,	*Clocks and more clocks*. Puffin, 1984.	**L,M,U**	8, 10, 83

Pigdon, K. & Woolley, M.,	*River Red.* Macmillan (Southern Cross), 1987.	**L,M**	68
Pomerantz, C.,	*One duck, another duck.* Greenwillow, 1984.	**L**	78
Prater, J.,	*On Friday something funny happened.* Random House, 1988.	**L**	
Randell, B.,	*How many legs?* Nelson (Windmill), 1985.	**L**	
Randell, B.,	*At the zoo.* Nelson (Joining-in books), 1986.	**L**	80
Randell, B.,	*Follow the leader.* Nelson (Windmill), 1985.	**L**	
Randell, B.,	*Houses.* Nelson (Windmill), 1985.	**L**	
Randell, B.,	*I've lost my gumboot.* Nelson (Windmill), 1986.	**L**	
Randell, B.,	*My tower.* Nelson (Windmill), 1986.	**L**	74-5
Randell, B.,	*Sea shells.* Nelson (Joining-in books), 1986.	**L,M**	72
Randell, B.,	*Ten little boats.* Ashton (Readalong Rhythm), 1985.	**L**	
Randell, B.,	*Ten little swimming crabs.* Nelson (Joining-in books), 1987.	**L**	80
Randell, B.,	*Ten small koalas.* Nelson (Joining-in books), 1986.	**L**	78
Randell, B.,	*What's the time, Mr Wolf?* Nelson (Windmill), 1985.	**L**	83
Reiss, J.,	*Shapes.* Macmillan, 1987.	**L,M**	
Reiss, J.,	*Numbers.* Macmillan, 1987.	**L,M**	79
Roennfeldt, R.,	**A day on The Avenue.** Penguin, 1983.	**M,U**	5, **39**, 83
Roffey, M.,	*Tinker, tailor, soldier, sailor.* Collins, 1979.	**L**	
Ryon, J.,	*Doodle's homework.* Collins, 1981.	**M**	
Scorfe, B.,	*And Billy went out to play.* Ashton Scholastic (Bookshelf stage 2), 1986.	**L**	
Serraillier, I.,	*The silver sword.* S. G. Phillips, 1959.	**U,PP**	
Smulleyman, R.,	*Alice in Puzzleland.* Penguin, 1982.	**U,PP**	
Smyth, G.,	**A pet for Mrs Arbuckle.** Crown, 1984.	**M,U**	23, **40**, 76-7
Stobbs, W.,	*The Derby ram.* Bodley Head, 1975.	**L**	68
Sutton, E.,	*My cat likes to hide in boxes.* Puffin, 1982.	**L,M,U**	74-5
Swih, J.,	*Gulliver's travels* (many editions)	**M,U**	68, 71
Tanner, J.,	*Niki's walk.* Macmillan (Southern Cross), 1987.	**M,U**	40, 76
Tarlton, J.,	*Danny's dilemma.* Ashton Scholastic, 1986.	**L,M**	25, 74-5
Tison, A.,	*Barbapapa.* Pan, 1973.	**L,M**	
Todd, H.E.,	*The crawly crawly caterpillar.* Carousel, 1981.	**L**	76
Tolstoy, A.,	*The great big enormous turnip.* Heinemann, 1974.	**L**	82
Traditional,	*Five little chicks.*	**L**	78
Traditional,	*Five little monkeys.*	**L**	24
Traditional,	**Goldilocks and the three bears.**	**L**	**10**, 20, 59, 69
Traditional,	*Jack and the beanstalk.*	**L**	68, 70
Traditional,	**King Kaid of India** (in *The Victorian Fifth Reader*, Ministry of Education, Victoria).	**U,PP**	2, 9, 46, **48**, 82
Traditional,	*Over in the meadow.*	**L**	79, 80
Traditional,	*Snow White and the seven dwarfs.*	**L**	
Traditional,	*The three billy goats Gruf.*	**L**	21
Traditional,	*The three little pigs.*	**L**	
Trinca, R.,	*One woolly wombat.* Omnibus, 1983.	**L**	
Viorst, J.,	**Alexander, who used to be rich last Sunday.** Macmillan, 1980.	**M**	**36**, 83
Viorst, J.,	*Alexander and the terrible horrible no good very bad day.* Macmillan, 1987.	**M**	37
Wheatley, N. & Rawlins, D.,	*My place.* Collins Dove, 1987.	**M,U**	53
Wild, M.,	**Something absolutely enormous.** Ashton Scholastic, 1984.	**M,U**	10, 12, **41**, 68, 87

Wildsmith, B.,	*All fall down.* OUP, 1987.	**L**	68
Wildsmith, B.,	*Animal shapes.* OUP, 1981.	**M,U**	74
Wildsmith, B.,	*Brian Wildsmith 123.* OUP, 1974.	**L,M**	74
Wildsmith, B.,	*The cat on the mat.* OUP, 1987.	**L**	68
Wildsmith, B.,	*Toot toot.* OUP, 1987.	**L**	79
Wildsmith, B.,	**The twelve days of of Christmas.** OUP, 1984.	**L,M**	**43**, 80
Williams, J.,	**The twelve days of Christmas.** Akers & Dorrington, 1984.	**L,M**	**43**
Williams, K.,	*Masquerade.* Schocken, 1977 (edition with solution, Workman, 1983).	**U,PP**	
Winer, Y.,	*Mr Brown's magnificent apple tree.* Ashton Scholastic, 1985.	**L**	57, 82
Wood, L.,	*Bump, bump, bump.* OUP, 1986.	**L**	
Zacharias, T. & W.,	*But where is the green parrot?* Doubleday, 1990.	**L**	74

Anthologies:

Count me in. Black, 1984

Davidson, A.	*Understanding Mathematics (set of seven books of number rhymes).* Shortland, 1984.	
Grice, M.	*One, two, three, four.* Warne, 1972.	
Haley, G.	*One, two, buckle my shoe.* World's Work, 1973.	
Martin, B.	*Sounds of numbers.* Holt, Rinehart & Winston, 1972.	68-9

Time for a number rhyme. Nelson, 1984.

Poems:

Behn, C.	'Circles', in *Rhymetime 2.* Arrow, 1984.	
Causley, **C.**	**'The song of the shapes', 'Twenty-four hours'**, in *Jock the treacle eater.* Macmillan, 1987.	40, 45, **63**, 74
Davidson, A.	**'Measuring'**, in *How many?* Shortland (Understanding Mathematics), 1984.	**55**, 68-9
	'Dogs', 'House shapes' 'How big is big?' 'Long' 'Round' 'Thick' 'Thin' 'Trees' 'What size are you?', in *How big is big?* Shortland (Understanding Mathematics), 1984.	68, 72-4
	'Six little firemen', in *What's the time Mr Wolf?* Shortland (Understanding Mathematics), 1984.	68-9
	'Buttons', in *Soldiers.* Shortland (Understanding Mathematics), 1984.	**19**, 72-3
	'Hands', 'Shapes' in *Ten little goblins.* Shortland (Understanding Mathematics), 1984.	**55**, 68-9, 74
	'Age', 'Apples', 'Autumn' in *Skittles.* Shortland (Understanding Mathematics), 1984.	**57, 62**
	'Band-Aids', 'Tangrams' in *Wheels.* Shortland (Understanding Mathematics), 1984.	74-5, 61
Farjeon, E.,	'Cottage', in *Rhymetime 2.* Arrow, 1984.	
Fyleman, R.,	**'A dreadful thought'** in *Wheels.* Shortland (Understanding Mathematics), 1984.	**58**
	'Mice' in Hilda Boswell's *Treasury of Poetry.* Collins.	**59**
Ireson, B.,	'At the supermarket', in *Rhymetime.* Hamlyn, 1979.	
Jacobs, L.B.,	'Fat old witch', in *Rhymetime 2.* Arrow, 1984.	
Kuskin, K.,	'Six times one', in *Rhymetime.* Hamlyn, 1979.	
Lehrer, T.,	'New maths', in *The Faber book of useful verse.* Faber and Faber, 1982.	
McGough, R.,	'Easy money', 'Money moans' in *Sky in the pie.* Puffin 1985.	37
Magee, W.,	'Down by the school gate' in *The Kingfisher book of comic verse.* Kingfisher Books, 1986.	

Mena, P.,	**'A spider's bedsocks',** in *The big bed*. Nelson (Ready to read), 1984.	21, **58**
Milne, A.A.,	'The end' in *Now we are six*, Dell, 1970.	68, 70
Patten, B.,	'Eight brand new angels', in *Gargling with jelly*. Puffin, 1986.	
Rosen, M.,	'The will', in *The Kingfisher book of children's poetry*. Kingfisher Books, 1985.	
Silverstein, S.,	'Channels', 'Eight balloons', 'How many how much', 'Shapes' in *A light in the attic*. Harper Junior Books, 1981.	
Silverstein, S.,	**'Band-Aids',** 'Eighteen flavours', 'The gypsies are coming', 'Hungry Mungry', 'Lester', 'Me and my giant', 'One inch tall', 'Smart' in *Where the sidewalk ends*. Harper Junior Books, 1984.	**60**, 87

References:

Bryant, M., *Riddles ancient and modern,* Hutchinson, London, 1983.

Carroll, L., *Alice's adventures in Wonderland,* in M. Gardner (ed.), *The annotated Alice,* New American Library, New York, 1974.

Clements, M.A., 'language factors in school mathematics', (1982), repr. in F*acets of Australian Mathematics Education,* Australian Association of Mathematics Teachers, 1984.

Dalton, J., *Adventures in thinking,* Nelson, Melbourne, 1985.

Donaldson, M., *Children's minds,* Norton, New York, 1979.

Egan, K., Teaching *as story telling,* U. of Chicago Press, Chicago, 1989.

Goodnow, J. and A. Burns, *Home and school: a child's eye view,* Allen and Unwin, Sydney, 1985.

Hughes, M., *Children and number,* Basil Blackwell, Oxford, 1986.

Ifrah, G., *From one to zero.* Penguin, New York, 1987.

Irons, R.R., 'Integrating language and mathematics to teach the concepts of the operations', in *Mathematics curriculum, teaching and learning,* Mothematical Association of Victoria, Melbourne, 1985.

Johnson, T., *Bringing it all together,* Heinemann, Portsmouth (N.H.), 1989.

Kline, M., *Mathematics in western culture,* OUP, New York, 1964.

Labinowicz, E., *Learning from children,* Addison Wesley, Menlo Park (California), 1985.

Mathematics Curriculum and Teaching Program, *Assessment alternatives in mathematics,* Curriculum Development Centre, Canberra, 1988.

Northrop, E.P., *Riddles in mathematics,* Penguin, London, 1963.

Wells, D., *The Penguin dictionary of curious and interesting numbers.* Penguin, London, 1986.

ACKNOWLEDGEMENTS

We should like to thank the many teachers and children from the following primary schools in Victoria who have helped us to develop and to trial the activities in this book: Antonio Park, Auburn, Belgrave South, Bimbadeen Heights, Birmingham, Camberwell, Cooinda, Glendal, Knoxfield, Macclesfield, Marlborough, Mooroolbark, St Mary's, Scoresby Heights, Stawell West, Strathbogie, Swan Hill, Syndal, Syndal South, The Basin, Upwey, Waverley North, Waverley Park and Wonga Park. Pam Hammond, Norma Pavey, Chris Trickey and Jenny Vincent have contributed ideas. Julie Fraser, Morwenna Griffiths, Andrew Herbert and Andrew Young took the photographs.

Parts of Chapters 1, 2, 3 and 7 have previously appeared in the following articles:

R. Griffiths, 'Picture books, stories and drawing for teaching mathematics', in P. Sullivan (ed.), *Mathematics curriculum, teaching and learning*, Mathematical Association of Victoria, Melbourne, 1985.

R. Griffiths and M. Clyne, 'Mathematics and children's literature', in N. Ellerton (ed.), *Mathematics: who needs what?*, Mathematical Association of Victoria, Melbourne, 1986.

R. Griffiths and M. Clyne, 'Literature in the mathematics classroom', in W. Caughey (ed.), *From now to the future*, Mathematical Association of Victoria, Melbourne, 1987.

R. Griffiths, 'Mathematics through children's literature', in J. Pegg (ed.), *Mathematical interfaces*, Australian Association of Mathematics Teachers, Armidale, 1988.

The quotations at the head of each chapter are from the 1974 edition of Norton Juster's *The phantom tollbooth* (Collins, London).

For permission to reproduce copyright material our thanks go to the following:

Avelyn Davidson for the poems 'Buttons', 'Hands', 'Age', 'Measuring', 'Autumn' and 'Apples' from the *Understanding Mathematics* series, Shortland Publications Ltd; the (UK) Society of Authors as the literary representative of the Estate of Rose Fyleman, for her poems 'A Dreadful Thought' and 'Mice'; the New Zealand Department of Education for the poem 'A Spider's Bedsocks' by Phil Mena, from *The Big Bed* in the *Ready to Read* series; Jonathan Cape Ltd for the poem 'Band-Aids' from *Where the Sidewalk Ends* by Shel Silverstein; David Higham Associates for the poem 'Song of the Shapes' by Charles Causley from *Jack the Treacle Eater* (Macmillan, UK); Collins Publishers for extracts from Norton Juster's *The Phantom Tollbooth*.